The 32nd Infantry "Red Arrow" Division in World War II

THE 32ND INFANTRY "RED ARROW" DIVISION IN WORLD WAR II

ROBERT YOUNG

Westphalia Press
An Imprint of the Policy Studies Organization
Washington, DC

The 32nd Infantry "Red Arrow" Division in World War II

Westphalia Press
An imprint of Policy Studies Organization
1367 Connecticut Avenue NW
Washington, D.C. 20036
info@ipsonet.org

ISBN: 978-1-63723-529-4

Daniel Gutierrez-Sandoval, Executive Director
PSO and Westphalia Press

Updated material and comments on this edition
can be found at the Westphalia Press website:
www.westphaliapress.org

TABLE OF CONTENTS

Author's Note: It will be noticeable that in some documents or quotes the 32nd of the 32nd Infantry Division is presented as 32d. In this text, it is always referred to as the 32nd Infantry Division.

LIST OF FIGURES

FOREWORD

World War II in the Pacific is best remembered as a naval war. Midway, Coral Sea, Leyte Gulf—some of history's most famous naval battles occurred during this war in the vast expanses of the Pacific Ocean between the United States and Japan. Even when one refers to land combat, the Marines receive the great majority of the recognition. The U.S. Army seems almost forgotten. The Army committed five times the number of men and suffered three times the casualties as did their Marine Corps brothers. No U.S. Army unit saw the Pacific on the scale of the 32nd Infantry Division. They were the first formation committed at Buna, the Army's first battle in the Pacific Theater of Operations (PTO). They also fought in its last campaign on the Philippine Island of Luzon. They began as an inexperienced collection of young and middle-aged American men from the upper Midwest. They ended the war tired and worn yet battle hardened and victorious. The success of this outstanding division often happened despite the failings of the organizations and leaders above them.

CHAPTER I

THE BACKGROUND OF THE PACIFIC WAR

While there had been several incidents and a decade of diplomatic back and forth between the United States and Japan, most of it over Japan's war in China, which began in 1931 when the Japanese invaded Manchuria, the Pacific War officially began with Japan's surprise attack against the U.S. Pacific Fleet's base at Pearl Harbor on 7 December 1941. The attack devastated the Pacific Fleet, but Japan's offensive extended throughout the Pacific and Southeast Asia. Japanese naval and ground forces quickly seized most of the Pacific Island chains. The American possessions of Guam and Wake were among the conquered territories. Burma, Indonesia, Malaya, Singapore, and parts of India all fell. The vast island archipelago of the Philippines, then an American Commonwealth, was also attacked on 7 December (the whole island chain would fall in May), several hours after the Japanese struck Pearl Harbor. The Philippines had a major airbase, Clark Field, on the island of Luzon, and an American army under the command of General Douglas MacArthur. That name and that commander became synonymous with the history of the 32nd Infantry Division.

MacArthur, a leader both praised and criticized by history and the men he commanded, did not distinguish himself in those first months of the war. Informed of the Japanese Pearl Harbor attack he did nothing to heighten the alert status of his command. Most of his air power was caught and destroyed on the ground. His army, poorly equipped and scattered throughout the islands, did not contest multiple Japanese landings. By the end of April 1942, the Japanese had conquered the islands and one of the most barbaric acts in the history of warfare occurred. The American Army surrendered and was force-marched from the Bataan Peninsula in southern Luzon to prison camps further north. Thousands of Americans and Filipinos died from abuse, starvation, and disease during the "Bataan Death March." General George Wainwright, the commander that surrendered the army to the Japanese, survived the march. Where was General MacArthur?

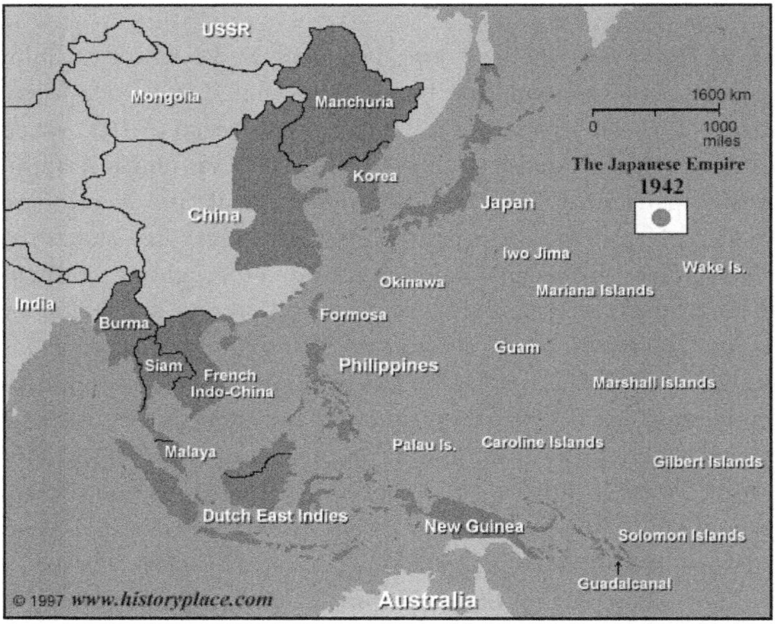

FIGURE I. As is clear above, the Japanese controlled an enormous amount of space. Much of it was in the Pacific, the rest along the Asian coastline. Southeast China and the Dutch East Indies provided most of Japan's vital resources. As an island nation Japan depended on the seaborne transport of these vital resources. The Philippines was in a position to interdict those supplies.

MacArthur, on the orders of President Roosevelt, had covertly left the Philippines with his family and key staff in PT Boats and gone to Australia. Looking for a hero in a time of constant depressing news, FDR nominated, and Congress approved the nation's highest military honor, the Medal of Honor, for MacArthur. The citation read:

> ***For conspicuous leadership in preparing the Philippine Islands to resist conquest, for gallantry and intrepidity above and beyond the call of duty in action against invading Japanese forces, and for the heroic conduct of defensive and offensive operations on the Bataan Peninsula. He mobilized, trained, and led an army which has received world acclaim for its gallant defense against a tremendous superiority of enemy forces in men and arms. His utter disregard of personal danger under heavy fire and aerial bombardment, his calm judgement in each crisis, inspired his troops, galvanized the spirit of resistance of the Filipino people, and confirmed the faith of the American people in their Armed Forces.***[1]

The political speechwriter that penned this citation deserved a Pulitzer Prize. Gallantry and intrepidity? Heroic conduct of defensive and offensive operations? Utter disregard of personal danger? Such grandiose rhetoric has never rung less true. Wainwright, who actually fought against the Japanese to the end was far more deserving of the nation's highest military award. MacArthur did utter a phrase before boarding his PT Boat, "I shall return!" The history of the 32nd Infantry Division and many more American units in World War II were created by that phrase.[2]

By early May 1942, Japan was the undeniable master of half the planet. Other than a militarily insignificant yet positive propaganda operation, 18 April's Doolittle Raid (Army medium bombers launched from aircraft carriers dropping a few bombs on a few Japanese cities) did convince the Japanese that their perimeter

---

1 Congressional Medal of Honor Society, https://www.cmohs.org/, Accessed 1 October 2023.

2 James Ellman, in his 2023 book *MacArthur Reconsidered: General Douglas MacArthur as a Wartime Commander*, obliterates the text of this citation in his examination of MacArthur's military record, in this case, with regards to the fall of the Philippines.

should extend further east and south, to the doorsteps of Hawaii and Australia. To the south they targeted New Guinea, the world's second largest island; if the Japanese could occupy Port Moresby, a key port in southern New Guinea, they could threaten or even attack Australia. To the east, the island of Midway, several hundred miles west of Hawaii, if occupied by the Japanese, threatened the Hawaiian Islands and the U.S. Pacific Fleet's main base. U.S. Navy cryptologists had broken Japan's main naval code and were aware of both operations. At the end of April, a naval/amphibious enemy force sailed for New Guinea, their objective Port Moresby. A U.S. task force centered on the aircraft carriers *Lexington* and *Yorktown* sailed to meet them. The result was the Battle of the Coral Sea.

The Battle of the Coral Sea is among history's most analyzed engagements. Losses were comparable. A Japanese light aircraft carrier, the *Shoho,* was sunk, and two large fleet carriers, the *Shokaku* and *Zuikaku,* were both damaged and put out of action for several months. American losses were also heavy. The *Lexington* was sunk and the *Yorktown* heavily damaged. History has called the battle a tactical draw, strategically, a U.S. victory. The Japanese invasion force turned for home. It was history's first naval battle where the opposing sides never fired a shot at each other. It was fought entirely by carrier-based aircraft. Several months later the Japanese did manage to land on New Guinea's northern coast and attempted to reach Port Moresby by a forced march through the thick jungle and imposing Owen Stanley mountains. They never reached Port Moresby.

The historic Battle of the Coral Sea was eclipsed by the titanic Battle of Midway. A massive Japanese naval armada, broken into three sections, sailed for Midway. The lead element had four fleet carriers, all Pearl Harbor veterans (*Akagi, Kaga, Hiryu, Soryu*), in addition to several dozen support ships. It would have had six fleet carriers, but the two carriers damaged at Coral Sea were several months away from rejoining the fleet. The second element had the amphibious group tasked with invading and securing Midway itself. The final element, the support element, contained numerous surface combatants including the new super battleships *Musashi* and *Yamato*. The top Japanese naval commander, Admiral Isoroku Yamamoto, the architect of Pearl Harbor and Japanese naval strategy thus far in the war accompanied the support element. He intended his lead element to knock out any planes on Midway's airfield and be in position to ambush the U.S. carriers in Hawaii as they sailed for Midway. The second element would by that time have attacked and secured the island. Yamamoto with his massive battleships would move in to finish off any American ships still afloat.

Initially, it was thought that only two U.S. aircraft carriers, the *Enterprise* and *Hornet,* would be available to oppose the Japanese. Due to some miraculous yard work the *Yorktown* was repaired and sailed in time to join her sister carriers northeast of Midway. The Japanese did not know they were there. The U.S. would ambush the Japanese.

At this point in the war Japanese naval aircraft and aircrews were far superior to their American counterparts. The three-day Battle of Midway, 4-6 June 1942, initially demonstrated this Japanese superiority. They eliminated the aircraft operating from Midway's airfield. U.S. patrol planes spotted the Japanese fleet and surprised them when the American carriers launched massed torpedo plane attacks. The torpedo planes were old, slow *Devastators.* Japan's superb fighter plane, the *Zero,* along with effective anti-aircraft fire, tore apart the *Devastator* squadrons. No hits were recorded. Despite these losses an opportunity emerged. Japanese planes, armed with bombs for a follow-up attack on Midway, were slowly and laboriously rearmed with torpedoes to attack the American carriers once they were found. Aircraft, fully fueled and armed, littered the Japanese carrier decks. Luck now entered the equation, and it went overwhelmingly to the Americans. A Japanese scout plane had spotted the American force, but radio trouble prevented it from giving the position. Japan's fighter cover had dropped predominantly to sea level because of the constant though futile torpedo attacks. U.S. *Dauntless* dive-bombers struck flight decks filled with aircraft. Three Japanese carriers were set afire and would later sink. The lone survivor, the *Hiryu,* down to less than two dozen planes, would launch two attacks; one with dive bombers, the other with torpedo planes, that knocked out the *Yorktown,* which would later sink. A final U.S. dive bomber attack destroyed the *Hiryu.* All four Japanese fleet carriers lay at the bottom of the Pacific. Their amphibious force, devoid of air cover, turned for home. Even their powerful support force, with the biggest battleships ever built, withdrew. The Battle of Midway was an unquestionable strategic and tactical American victory. In addition to their ship losses, the loss of irreplaceable aircrews would cripple the Japanese for the rest of the war.

The Japanese would launch no more major offensives in the Pacific. However, they did consolidate what they had already conquered. Consolidation in most cases involved construction of defensive positions and airfields. One area where this consolidation particularly alarmed American military leadership was the Solomon Islands, specifically the island of Guadalcanal. An operational airfield on Guadalcanal threatened America's lines of communication and supply with Australia and possibly even Australia itself. On 7 August 1942, only eight months after the devastation of Pearl Harbor, America launched its first offensive operation of World War II—the invasion of Guadalcanal.

The 1st Marine Division landed on Guadalcanal and moved towards the island's airfield, later famously named Henderson Field, and a six-month campaign followed. Bitter fighting, both on land and at sea, signified the battle. What developed into a campaign of attrition ended in an American victory. The fighting at Guadalcanal dominated the headlines in America. Where was General MacArthur?

After fleeing the Philippines MacArthur established a headquarters in Australia for what became the Southwest Pacific Area (SWPA). New Guinea gave him the opportunity to enter the war. Japan's first attempt to reach Port Moresby was thwarted at Coral Sea. On 22 July they landed at Gona, on New Guinea's northern coast, marched to Buna, and began to march towards Port Moresby, one hundred miles away. The distance may not seem excessive, but the advance was along a single, treacherous track, the Kokoda Trail, and through high mountains. Veteran Australian infantry fought against a force of some 5,000 and were pushed to within 20 miles of Port Moresby. They went no further. Horrible terrain, constant rain, supply problems, and Allied air attacks drove them back towards Buna. For the next few months, the Japanese would sporadically land troops at Buna, often unbeknownst to American intelligence, while what remained of the 5,000 men that had marched towards Port Moresby lost most of their number to exhaustion, disease, and starvation. Those that did make it back were no longer battle worthy. These were the men MacArthur was told remained at Buna. Two valuable airstrips were also there. MacArthur had two American infantry divisions in Australia, the 32nd and 41st. The 41st was still arriving. MacArthur ordered the 32nd to Buna.[3]

MacArthur would dominate this theater of the war and at this time was probably the nation's most famous military figure. From a distinguished military family his father had been a Union general during the Civil War and in a heralded career even rose to Military Governor of the Philippines. He never reached the top army post, Chief of Staff. MacArthur graduated from West Point in 1903 at the top of his class and served on the front lines in World War I as the Assistant Division Commander of the 42nd Infantry Division. He was decorated for gallantry and at war's end became the Division Commander. From there he became Superintendent of West Point, and by the end of the 1920s the Army Chief of Staff. His time as Chief of Staff was filled with problems and controversy. He oversaw the court-martial of air power advocate General Billy Mitchell and ordered the "clearing" of the Bonus Army, World War

3 I Corps, *History of the Buna Campaign: December 1, 1942 – January 25, 1942,* (Headquarters: I Corps, 1943), 12-14.

I veterans seeking bonus payments from Washington. MacArthur also had to deal with dwindling budgets, numbers, and a nation focused on isolationism. He spent many years as a child in the Philippines and now returned there after finishing his time as Chief of Staff as the military advisor to the Filipino Governor. This new position forced him to leave the army. Since the end of World War I, MacArthur seemed to detach himself from the man who led from the front and cared so deeply for the men he led during WWI.[4] His personal glory and opinions would dominate the rest of his public and military career. In the months before America entered World War II he returned to the army and his four-star rank. He did have capable subordinates. Those capable commanders were in the field, not his personal staff. At Buna the main subordinate was Lieutenant General Robert Eichelberger.

Robert Eichelberger graduated in the West Point class of 1909. He served in Panama and the Mexican Expedition against Pancho Villa but not in World War I. He earned kudos and the Distinguished Service Cross, America's second highest military honor, for his service with the American Expeditionary Force in Siberia during the Russian Civil War. Eichelberger then served in China and the Philippines, attended the Army's General Staff and War Colleges, and in 1940 was promoted to Brigadier General and then became the Superintendent of West Point. In early 1942 he left the Academy and returned to the active army. MacArthur knew him well and trusted him to evaluate the 32nd Infantry Division during the summer of 1942 and then institute a rigorous training program. Unfortunately, MacArthur ignored his recommendations and did not allow his training program the time it needed to effect change. Eichelberger entered the Buna campaign later and would earn the nickname "MacArthur's Fireman" for the many times he saved his boss from himself.

---

4 James Ellman, *MacArthur Reconsidered: General Douglas MacArthur as a Wartime Commander*, (New York: Stackpole Books, 2023), 11.

FIGURE II. General Douglas Macarthur and his "Fireman," General Robert Eichelberger.

FIGURE III. General Edwin Harding.

General Edwin Harding commanded the 32nd Infantry Division for most of 1942. A West Point classmate of Eichelberger, he also remained in the United States during WWI. He served in China and various staff positions until the mid-1930s where he was first, the Secretary of the U.S. Infantry Association and then the Editor of the Infantry Journal. In 1938 he was given command of the 27th Infantry Regiment, then promoted to General and sent to the 9th Infantry Division as its Assistant Division Commander. He was given the 32nd Infantry Division in February 1942.[5]

General Edwin Harding led the 32nd to Buna. He did not command the division when the war began; Major General Irving Fish, a veteran of the 32nd's time at the Mexican border in 1916 and in World War I, was relieved of command in February 1942. Harding replaced him due to a then recently instituted "over-age" policy that the Division neither liked nor found fair. The 32nd's Veterans Organization makes this clear:

> In truth, the creation and enforcement of this 'over-age' policy was little more than a thinly veiled excuse to get rid of senior National Guard officers and give their desirable commands to Regular Army officers. To make a long, complicated story short, the Regular Army in 1940 was bloated with officers, especially colonels, but other officer ranks as well. This excess in officers was partly caused by the fact that the strength of the Regular Army was drastically reduced after World War I; they got rid of many enlisted soldiers but kept many officers. Also, the promotion system for officers between the wars was very ineffective. Sometimes the Regular Army officers that replaced the so-called 'over-age' National Guard officers were themselves over-age or later became over-age but

5 https://mwi.westpoint.edu/harding-project-armyauthor-profile-major-general-edwin-forrest-harding/, Accessed 1 February 2024.

> were not replaced when they did. Another tactic used to replace National Guard officers with Regular Army officers was to give the National Guard extremely rigorous physical examinations, much more thorough than those given to enlisted soldiers, junior officers or Regular Army officers. In this way they could create more vacancies for Regular Army officers by claiming that some of these National Guard officers suffered from other unnamed or nonexistent ailments. This information is not being included here to imply that these Regular Army officers were in any way undeserving or unqualified for these positions, it is only being included to point out that some National Guard officers were treated in an unjust and unprofessional manner by some in the Regular Army.[6]

Episodes like the above are not uncommon in American or any other nation's military history. National Guard divisions like the 32nd serve many years together, the transfers and rotations that exist within Regular units are rare. Leaders and the men they command know each other. This 'over-age' system likely created animosity and an unwelcome, unnecessary distraction.

General Harding led a division that never seemed to be in one place for any extended period. Originally destined for Ireland and then Europe, the Division was notified in March it was now heading for Australia. The 32nd, on paper, looked like this:

Military Police Company 126TH Infantry
Regiment 127TH Infantry Regiment 128TH Infantry Regiment

Division Artillery Headquarters and Headquarters Battery (formerly HHB, 57TH FA Bde.)

120TH Field Artillery Battalion, Lt. Col. Harold A. Morgan, from Merrill, WI

6 https://www.32nd-division.org/, Accessed 1 February 2024.

121ST Field Artillery Battalion, Lt. Col. Melvin Leslie McCreary, from Milwaukee, WI

126TH Field Artillery Battalion, Lt. Col. Ross J. Quatsoe, from De Pere, WI

129TH Field Artillery Battalion, Col. Ralph A. Loveland, from Ann Arbor, MI 107TH Medical Battalion, Lt. Col. Carl Hanna, from Detroit, MI 107TH Quartermaster Battalion, Lt. Col. Donald M. Farris, from Madison, WI 32D Signal Company 32D Cavalry Reconnaissance Troop (formerly HQ, 64TH Inf. Bde.) 632D Tank Destroyer Battalion (not organic to the Division), Maj. Roy William Bailey, from Eau Claire, WI[7]

Of the above the three infantry regiments, parts of the 107th Medical and Quartermaster Battalions, and a single 105-mm gun from one of the artillery battalions went to Buna. No tanks, no Cav Troop, and only the ONE gun. Most of the Division's attachments and non-infantry men did not arrive in Australia until late in the year. When the 32nd arrived in San Francisco they received 3,000 replacements, all basic trainees. They left San Francisco on 22 April, arrived in Australia on 14 May, made camp, and then moved again in July to a new camp near Brisbane. A month later they began moving to New Guinea.[8]

7 Ibid.

8 Samuel Milner, *Victory in Papua,* (Washington D.C.: Office of the Chief of Military History, 1957), 112-115.

CHAPTER II

BAPTISM OF FIRE

> Unhappiest of November's hard realization was the plight of the Buna ground forces. In this, their grueling baptism of fire, they found the Japanese in perfect defensive positions. The Japanese left flank was based on the tidal, unfordable Girua River and Entrance Creek; their right flank was based on the sea, as was their rear. To their front was think jungle bogged into deep swamps traversed by tracks which canalized any American attack into narrow fronts against prepared positions.
>
> —I Corps, *History of the Buna Campaign*

If one were looking for the ideal place to fight a battle the last place they might look is Buna. The following is a detailed analysis, produced after the battle of course, of the natural geography of Buna, created by the Military Intelligence Division of the War Department:

> The enemy had taken cunning advantage of the peculiar geography of the Buna area to construct two almost impregnable defensive lines hidden in the swampy jungle. One of those lay across the Soputa-Sananda Road; the other was in front of Buna itself. Our failures on 18 November had only the vaguest notion of these fortifications.
>
> The fighting took place in a narrow strip on the northeastern coast of Papua extending on both sides of Buna Mission, the prewar seat of government for the area. The Mission, sometimes called

the Government Station, consisted of three houses and a few dozen native huts. Buna Village, a half mile to the northwest was merely a cluster of huts. Bounded on the west by one of the several outlets of the Girua River and on the east by the coast a half mile south of Cape Endaiadere, the battleground was about 3.5 miles long and ¾ of a mile deep.

The principle swamp in he Buna area lies between Entrance Creek and Simemi Creek and reaches inland to the vicinity of Simemi and Ango. It is absolutely impenetrable, a fact of vital importance in the campaign. Between the closely spaced trees, which are 25 to 100 feet high, is a tangle of roots, creepers, and underbrush. A man standing up can see from 5 to 30 yards; from a foxhole visibility is practically zero. Much of the other ground in the area, though not actually swamp, is thoroughly waterlogged, but a few places near the shore are fairly dry.

Most of the drier land is covered with a thick growth of Kunai grass or plantations of coconut palms. This coarse grass grows to a height of more than 6 feet, but its height varies greatly, depending on how recently it has been burned over or cut. Its leaves are broad and sharp-edged; its stems are about the thickness of a pencil.

Approach to Buna is difficult whether by sea or by land. It has no harbor. Coral reefs abound near the shore and are scattered over the sea to a distance of 25 miles from land. Cargo has to be discharged at sea into native double canoes usually carrying 1 to 1.5 tons. On the landside, Buna is cut off by swamps

> and creeks and can be approached through only four narrow corridors, each with its trail.[8]

The terrain was not the only problem. It was hot and humid. Torrential rains could appear out of nowhere. The ground was rampant with disease. The troops lacked basic equipment necessary for this environment, including machetes and insect repellant.[9] Of course neither MacArthur nor his Staff were aware of any of this.

That is not surprising since a source affirmed: "... no senior officer had come forward along the Kokoda Trail farther than where the motor road ended ... about ten miles short of the point of the farthest Japanese advance."[10] These were men predominantly from the Midwest. The mountains, jungles, heat, insects, noises, etc., all were completely alien to the soldiers of the 32nd Division. Had they undergone a rigorous training and acclimation program before entering combat the problems encountered with the environment could have been mitigated. What exactly was the state of the 32nd on the eve of battle?

General Harding offered an opinion on the above question after the war. His division and men were criticized for their lack of stamina and combat readiness. The General pointed out:

> I have no quarrel with the general thesis that the 32d was by no means adequately trained for combat—particularly jungle combat. From February when I took over until November when we went into battle, we were always getting ready to move, on the move or getting settled after a move. No

8 Military Intelligence Division, *Papuan Campaign: The Buna-Sananda Operation, 16 November 1942 – 23 January 1943,* (Washington D.C.: War Department, Military Intelligence Division, 1944), 9-11.

9 *Lieutenant Colonel Laurence M. Jones, Defense of the Philippines to the Battle of Buna: A Critical Analysis of General Douglas MacArthur,* (Maxwell Air Force Base, Alabama: Air War College, 1989), 42.

10 Ibid., 43.

sooner would we get a systematic training program started than orders for a move came along to disrupt it.[11]

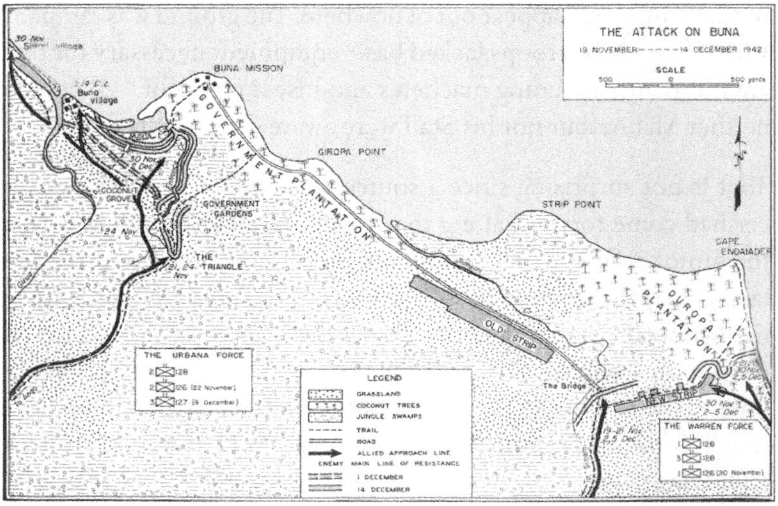

FIGURE IV. The Attack on Buna

Above is the Buna battlefield. Note the key airstrips, separated by a river with the only passage point the small footbridge. All the area surrounding the key terrain was swampy jungle or grassland. The entire battlefield was only a mile wide but due to terrain restrictions it may as well have been ten times that. An amphibious landing could have secured any of the key points but Japanese guns, a lack of water transport and any naval gunfire support, and no experience or training in amphibious tactics made such a landing impossible.

General Harding believed, that given time, the jungle-training program he initiated would have worked.[12] General Eichelberger, the soon to be commander of the I Corps, the American parent unit of the 32nd and 41st Infantry Divisions, observed the 32nd in the

11 Milner, *Victory in Papua*, 133.

12 *Samuel Milner Interview with General Harding*, 9 December 1947.

months before fighting began. He reported these observations to General MacArthur, noting:

> In Washington I had read General MacArthur's estimates of his two infantry divisions and these reports and our own inspections had convinced my staff and me that the American troops were in no sense ready for jungle warfare. It was true that we were newcomers, but I had expected to learn lessons in jungle training in Australia. And it seemed to me that our troops in training were just being given more of the same thing they had done back home ... I told General MacArthur and Sutherland (MacArthur's Chief of Staff) that I thought the 32d Division was not sufficiently trained to meet Japanese veterans on equal terms. After an inspection—and this must be in the War Department records somewhere—I gave the 32d Division a "barely satisfactory" rating in combat efficiency.[13]

Eichelberger instituted a jungle-training program but as with any type of training General Harding tried to begin movements prevented its implementation. This shortcoming revealed itself when a battalion of the 128th Infantry Regiment (IR) marched to the Buna area from the south, over the Owen Stanley Mountains, a route that previously ravaged a far more seasoned Japanese force.

The War Department dispatched observers to all areas of the war where American forces were engaged. One such observer noted that the lack of preparedness amongst American infantry officers and riflemen went beyond physical conditioning. Tactically, they had no idea how to operate in the jungle. All that was taught in the service schools in the United States did not apply in New Guinea. One observer noted: "The application of tactical principles used

13 Robert Eichelberger, *Our Jungle Road to Tokyo,* (New York: The Viking Press, 1947), 29.

in jungle warfare in New Guinea is radically different from those taught in our service schools. There is practically no distribution in depth and little chance for maneuvering except infiltration by patrols. Very small reserves were held out and those were close to the front line."[14]

A seaborne landing at Buna was risky and the area lacked the shipping necessary to execute such a landing. The 126th IR flew to a point where the drive on Buna could commence. The 127th arrived in the middle of December. The 128th walked. An officer accompanying this march continuously recorded the poor condition of the American soldiers. His comments, attached to the Division's Buna After Action Report (AAR), should have warned those planning the Buna operation the in-house problems awaiting them.

The Observer's negative comments began on the first day. He noted, "... while the trail covered during the first day of travel was not particularly difficult, the fact was disclosed that the personnel in the column were in poor condition. Considerable straggling developed and much time was lost ..."[15] The second day was even worse, and the observer's comments reveal not only overall unit readiness but his own personal physical stamina:

> The second day's march proved difficult for a large portion of the column. It disclosed the information that many men started who should have remained behind. Stragglers were suffering from dysentery, some just recovering; others were groups of men who had done very little marching, prior to starting out. This column had in it 30 personnel from hospital unit, personnel from Hq. Co., 107 Med. Det., including a number of officers

14 Army Ground Forces, *Observers Reports w/Appendixes: New Guinea*, (Headquarters: Army Ground Forces, 1943), 3.

15 32nd Infantry Division, *After Action Report: Buna*, (Headquarters: 32nd Infantry Division, 1943), 1.

> who had experienced little or no marching. I was included in this group, finding it extremely difficult to keep going due to the fact in two years of service I had not marched over two miles at any one time ... I had also observed Capt. Harper, Bn S-3, having difficulty along the trail. I was having difficulty myself, though I managed to move at a fair rate. Upon arrival at bivouac area, somewhat exhausted, I reported to Col. Geerds. I discovered that he had suffered two heart attacks on the trail during the day ... I also learned that Capt. Laughter, Hq. Co. CO, was exhausted and shortly after Captain Gutman and Captain Harper arrived ill and exhausted. Also that Lt. Carter, Company F, and a few enlisted men from preceding units were ill at this carrier station. A check of the troops condition in general disclosed that over half of the column was nearly exhausted.[16]

It did not stop with the fact so many were unable to execute a grueling march. Most, as the observer noted (he himself had NEVER marched over two miles at any one time), could not execute even a basic march. Men were not conserving water, drinking all they had in the first hours of the moves.[17] Cold also became a factor. New Guinea is a tropical, jungle covered land. In the high elevations of the Kokoda Trail and Owen Stanley Mountains, the temperatures drastically dropped and cold rains often drenched the troops. One Midwestern officer noted:

> ... The terrain was ghostlike in form with the whole atmosphere weird and fantastic. The conditions existing in crossing the top of this mountain defies proper description. We kept on and on and even-

16 Ibid., 2.

17 Ibid., 3.

> tually started down on the fast side. The cold was so intense and everything so completely soaked that stopping for bivouac was out of question until considerable descent had been made ... Troops were soaked, cold, hungry, and exhausted.[18]

When finally arriving at their destination, the unit was debilitated, unable to enter combat for several weeks. The observer's report concluded with several recommendations which showed not only the failure of not instituting a jungle training program but also the inexperience of staff sections with regards to supply and administration. Major Garmet J. Burlingam (the observer) recommended:

1. That all personnel assigned a mission calling for extreme physical exertion such as given the 2nd Bn be picked men as to physique, physical condition, tactical experiences, past record, and mental attitudes.
2. That no officer over the age be assigned to accompany a column on a march of this nature.
3. That the prescribed ration of two C cans and rice per day be increased to minimum of four per day with two chocolate bars and rice. This trip proved without question that the prescribed ration for trail travel was very inadequate to replace the physical energy expended.[19]

Units with the personnel limitations and administrative failings of the 2nd Battalion, 128th Infantry Regiment were clearly not ready for combat. The 32nd Infantry Division was drowning in these limitations though they never had the chance to correct them in a rigorous training program. General Eichelberger warned MacArthur of the 32nd's limitations. Despite these doubts, Eichelberger approved sending this battalion over those mountains.[20] General

18 Ibid.

19 Ibid., 11.

20 Unknown Author, *Eclectic Memories and Opinions, 32nd Infantry Division, WWII, Southwest Pacific,* (Wisconsin: 32nd Veterans Association, 1946), 5.

Harding recognized them as well. MacArthur never personally inspected the Division, either before committing them to action nor during the battle itself. Yet in his *Reminiscences,* written after the war, he points out these deficiencies, no doubt to glorify himself as securing victory despite them, he extolled:

> None of the three elements of my command—naval, air, or ground—was adequate ... The ground troops too, were inadequate for the tremendous task ahead. Not only were they too few in number, but also they lacked the equipment and the strenuous training necessary for combat. The shortage of water transportation for supplies and equipment, as well as for troops, was one of the most difficult problems of all. Sufficient shipping was unavailable to adequately mount operations and then to support the troops in widely dispersed locations.[21]

Then why did MacArthur send this division to action? He only at that time had two American divisions available, the 32nd and 41st. The Japanese airfields, one complete, the other in process, were a threat. Further, it was thought that only the tattered Japanese survivors of their own harrowing march through the Owen Stanley's defended Buna. That was an enormous intelligence failure.

Battlefield intelligence is of course crucial to success. Prisoner interrogations, vigorous patrolling, local sources, all are vital tools to an attacking force. America had an additional asset, *ULTRA*. *ULTRA* was the most closely guarded secret of the entire war. British and American cryptologists broke both the German and Japanese codes, allowing them to read **ALL** the radio traffic on military circuits. Unit locations, strengths, weaknesses, gripes, etc., anything on the enemy radios went to Allied commanders. Not all commanders. It usually did not go below army level. Corps and division

21 Douglas MacArthur, *Reminiscences,* (New York: Time, Inc.: 1964), 156.

commanders never saw it. Yet for all *ULTRA's* value, especially later in the war, it did little here. General George Kenney, commander of MacArthur's Fifth Air Force, believed his planes could as well as being the "artillery of the jungle" be the chief reconnaissance asset. A few weeks in the jungle by an Army observer dispelled that notion. He noted that "Air Reconnaissance was of little use in jungle warfare and practically all combat information had to be obtained by ground reconnaissance and the use of patrols."[22] The problem once the battle began was few patrols happened. Before the first attack NO patrols in Buna ever occurred.

Despite their own deplorable condition, the men of the 32nd believed Buna would be easy. The intelligence failures started at the top. Eichelberger noted:

> ... Allied headquarters originally expected that Buna would be taken without opposition. But fresh troops had been landed there, most of them veterans of the Jap campaigns in China, Malaya, and the Philippines. Actual fighting at Buna started on November 20. So ill-informed was the Intelligence Section of our 32d Division that enemy strength east of the Girua River was estimated at only three hundred when there were about three thousand men there.[23]

Samuel Milner, author of *Victory In Papua,* the Army's official history of the campaign, noted that despite all their many problems the troops were "cocky and overconfident" and believed Buna would be a "pushover."[24] That overconfidence extended to General Harding, who stated: "I think it is quite possible (Oct. 14) that the Japanese may have pulled out some of their Buna forces ... we

22 Army Ground Forces, *Observers Reports w/Appendixes: New Guinea,* (Headquarters: Army Ground Forces, 1943), 4.

23 Eichelberger, *Our Jungle Road to Tokyo,* 36.

24 Milner, *Victory in Papua,* 137.

might find Buna easy pickings with only a shell of sacrifice troops left behind to defend it."[25] Even a Ground Forces observer dispatched by the War Department, Colonel H.F. Handy, noted that many of the men of the 32nd felt "Buna could be had by walking in and taking over."[26] They, and tragically their higher headquarters, did not know of the reinforcements the Japanese landed.

There was no knowledge of the strength of Japanese positions. Between Buna and Gona there were 8,000 enemy troops. Intelligence placed that number at 1,500-2,000.[27] Japanese convoys landed reinforcements throughout August and September. Aerial reconnaissance saw nothing. *ULTRA* had no idea. As historian Edward Drea, the leading historical authority on *ULTRA* noted: "Neither ULTRA nor traditional battlefield intelligence had penetrated the cloak of secrecy that surrounded Japanese operations and obscured their deployments." [28] MacArthur's incompetent Chief of Intelligence, Colonel Charles Willoughby, did not help the situation. He, as did many others, underestimated Japanese strength. His greatest error, as Drea pointed out, involved misreading Japanese intentions, declaring "further reinforcement (after August) of Buna was improbable because of Japanese commitments in the Solomons, logistics difficulties, and the high risks of such a gamble ... Instead, the Japanese response to the American landing at Oro Bay was as swift as it was violent."[29] This overconfidence extended to the weapons or lack thereof given the 32nd before entering battle.

An American infantry division possessed a uniform allotment of weapons. There were four artillery battalions, three of 105-mm. howitzers and one of 155-mm. howitzers. Each battalion had

25 Ibid.

26 Ibid., 138.

27 *Lieutenant Colonel Laurence M. Jones, Defense of the Philippines to The Battle of Buna: A Critical Analysis of General Douglas MacArthur*, 43.

28 Edward Drea, *MacArthur's ULTRA: Codebreaking and the War against Japan, 1942–1945*, (Kansas: University of Kansas Press, 1992), 51.

29 Ibid.

twelve guns. This has already been stated but must be repeated, the 32nd brought ONE gun with it to Buna. They left 2/3 of their 81-mm mortars behind.[30] Lack of transport was the main reason though given the belief of how little awaited to oppose them, no sense of urgency existed to get the rest of the division artillery forward. General Kenney also played a role here. Kenney called aircraft "the artillery of the jungle."[31] He doubted artillery's effectiveness, due to lack of observation, in the heavy foliage of a jungle. At this early point of the war, his pilots and planes were no better, and air power did little to affect the Buna battlefield. Shortages did not stop with artillery and mortars. They were short of everything. They were on 1/3 rations as they moved forward.[32] Unconditioned men expending calories in the heat and humidity of a jungle environment, even in a less than strenuous combat situation, would quickly be malnourished on such short rations. The lack of artillery was a major concern of General Harding and General Waldron, the Division Artillery Officer. Harding noted, "... we were told the Australians had plenty of guns in New Guinea. He (Waldron) and I wanted all the arty we could get."[33] Intelligence failures helped create this situation and failed again when completely missing just how strong the positions around Buna were.

The Japanese did an outstanding job creating bunkers that commanded all the key approaches to Buna's most critical points:

> The base of the bunker was a shallow trench, up to 40 feet in length for the large bunkers, and 6 to 10 feet for the smaller. A framework of columns and beams was set up, the walls were riveted with

30 Military Intelligence Division, *Papuan Campaign: The Buna-Sananda Operation, 16 November 1942 – 23 January 1943*, 4.

31 George Kenney, *George Kenney Reports*, (New York: Duell, Sloan, and Pearce, 1949), 140.

32 *Lieutenant Colonel Laurence M. Jones, Defense of the Philippines to the Battle of Buna: A Critical Analysis of General Douglas MacArthur*, 43.

33 Samuel Milner, *Interview with General Harding*, 9 December 1947.

> coconut logs ranging up to 1.5 feet in thickness, and a ceiling of two or three courses of such logs were laid on top. Not content with this construction, the enemy reinforced the wall, using steel oil drums and ammunition boxes filled with sand, as well as log piles and rocks. Over all this were piled earth and sand mixed with short logs, coconuts, and the like. When the bunker, 7 to 8 feet high, was camouflaged with fast growing jungle vegetation, it became almost impossible to spot in the tangled underbrush.[34]

This is what faced the 126th and 128th Infantry Regiments; the 127th would not arrive for another month, as they launched their first attack against Buna on 19 November 1942.

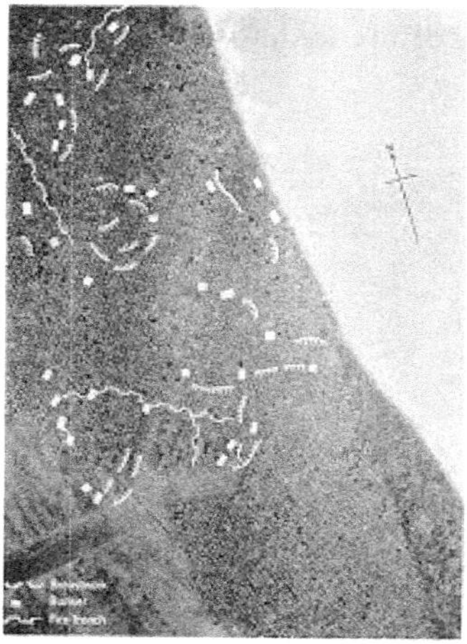

FIGURE V. Japanese Defenses in the Duropa Plantation

This, just one part of the battlefield, displays the vast network of Japanese defenses. At this point alone there were 32 bunkers, 28 fire trenches, and several networks of breastworks. Attacks would be channeled to the weapons of the bunkers, all expertly camouflaged.

34 Military Intelligence Division, *Papuan Campaign: The Buna-Sanananda Operation, 16 November 1942 – 23 January 1943*, 4.

FIGURE VI. The Bridge over Simemi Creek

This small bridge, more realistically just a small footpath, connecting the Old and New Strips. Note of course the incessant jungle and the vulnerability of any soldiers crossing to fire.

FIGURE VII. The Interior of a Japanese Bunker

Note the oil drums, all sand filled, protecting the walls of the bunker and the coconut logs bracing it.

FIGURE VIII. The Exterior of a Japanese Bunker

Note the expert camouflage. Almost impossible to locate unless right on top of it or after being fired upon.

The first attack of 19 November was expected to go rather easily. General Harding had hoped to precede the attack with an artillery barrage and air attack. The night and morning saw heavy rains, grounding all aircraft and soaking the men of the 128th IR, the attacking force. Harding had hoped to have two regiments in this first attack but a few days earlier the 126th IR was given to the Australians a few miles away at Gona. Though two battalions of the 126th would return a few days later their absence in these first attacks was significant. The artillery available was the one American howitzer and two Australian 25 pounders (3.7 inch). A quick bombardment announced the attack of the 128th IR's 1st and 3rd Battalions.

The 1st Bn moved forward, their objective Cape Endaiadere and the sea about two miles past their start point. To get there they had to pass through the Duropa Coconut Plantation. This was supposed to be easy (Refer to Figure IV). A few hungry, tattered enemy soldiers. There was no need of reconnaissance or patrols to find non-existent positions. The shock must have been incredible when from seemingly every direction the 1st Bn's troops were engulfed by Japanese machine gun fire. The thick jungle masked sound and the enemy weapons were flashless.[35] The attack quickly disintegrated, and the U.S. troops pulled back. An Engineer officer, attached to the 1st Bn, observed the attack and the problems encountered:

> The first opposition from the enemy here was a surprise and shock to our green troops. The enemy positions were amazingly well camouflaged, and seemed to have excellent fields of fire even in the close quarters of the jungle ... Snipers were everywhere ... and they were as well camouflaged that it was nearly impossible to discern them. The enemy habitually allowed our troops to advance to very close range—sometimes four or five feet from a machine gun post—before opening fire; often they allowed troops to bypass them completely, opening fire then on our elements, and on our front elements from the rear.

He added:

> ... Our troops were pinned down everywhere by extremely effective fire. It was dangerous to show even a finger from behind one's cover, as it would immediately draw a burst of fire. It was impossible to see where the enemy fire was coming from; consequently our own rifle and machine gun fire was ineffective during the early stages ... Grenades and mortars were difficult to use because, first, it was difficult to pick out a nest position to advance upon with grenades, second, the thick jungle growth and high grass made throwing and firing difficult, and third, because it was nearly impos-

35 Milner, *Victory in Papua*, 175.

sible to observe our fire.[36]

The 1st Bn troops had no choice but to fall back. They did not know what to shoot at. As the campaign progressed the exact locations of the Japanese positions were located. Even when located problems still persisted as will be seen in the coming pages.

The 3rd Bn fared no better. They advanced up the Simemi Trail towards the Bridge, the main junction between the Old and New Strips. The area was swamp-infested jungle except for some dry land approximately 300 yards south of the Bridge. Once they got to that dry land machine gun fire engulfed them as it had the 1st Bn. No advance was possible. As the Battalion Commander, Colonel Miller, honestly stated: "We were stopped cold!"[37] The 19th saw no further action from the 3rd Bn and they could not attack on the 20th. An air attack by some medium bombers (A-20s and B-25s) and a barrage by 60-mm mortars and a few of the Australian 25 –pounders did little. U.S. troops, with an idea now of their locations, eliminated several enemy machine gun nests. They advanced several hundred yards before fire from bunkers and other machine guns stopped them.

On 21 November two battalions, one from the 126th IR, the other from the 128th IR, attacked. An air strike again preceded the main attack. More successful than the previous day's strike, it hit a few Japanese machine gun nests. However, the infantry did not immediately follow the attack and no artillery fired. The attack failed and though the air strike hit Japanese positions, some of it once again fell short and hit American positions. The infantry had not attacked because the battalion commanders did not receive their orders to attack until 0840 and 0850 hours, forty and fifty minutes after the air strike.[38]

36 Ibid.

37 Ibid., 176.

38 Ibid., 177.

This day would reveal a great deal about the support available to the 32nd ID during this initial phase of the campaign. General Harding wanted another air strike in the afternoon before launching an attack. He, after much cajoling, got the air commanders to launch another strike. Once again it was very inaccurate and once again it hit American troops, killing six and wounding twelve.[39] The infantry would finally go forward late in the afternoon after a brief, unobserved barrage by the few Australian guns and the available mortars. It had no effect on the Japanese positions and the attack was easily thrown back.[40]

These first attacks demonstrated how ineffectual air power was in a jungle environment. The Japanese positions could not be spotted from the air. Other than flying at treetop level there was also little chance of an accurate, pinpoint attack on a specific bunker or position.[41] General Kenney's boast that air power was the artillery of the jungle rang very shallow throughout the Pacific War. Artillery also did little without a direct hit, an impossible feat when position locations were unknown and the fire unobserved. Artillery ammunition, at least what was available at Buna, was also ineffective. Japanese pillboxes were thick and covered in jungle vegetation. The ammunition was quick fused, exploding on contact. All it did was blow around the overhead cover. Delayed action fuses were needed to penetrate the covering foliage before exploding. General Harding noted all the above and realized he needed tanks, which could be used on the dry ground of the Duropa Plantation.[42] Three tanks, light M-3 Stuarts, were put on a captured Japanese barge for movement to Buna. As soon as pushing off the barge and tanks sank. Harding's repeated future requests for reinforcements were either ignored or refused.

---

39 Ibid., 178.

40 Ibid.

41 32nd Infantry Division, *After Action Report: Buna*, 7.

42 Milner, *Victory in Papua*, 179.

These first failures also drew the ire of General MacArthur. Though he NEVER visited the front during the entire campaign he began criticizing Harding and his men for both a lack of success and a lack of an aggressive spirit. The 32nd's porous supply situation and his own staff's intelligence failures concerning Japanese strength and dispositions never seemed to enter his mindset. Inexperience was an enormous problem here. These men did not yet know how to be soldiers. An Observer noted this and the reticence inexperienced American soldiers felt in the jungle was all too obvious:

> During my observations on New Guinea I was constantly impressed by the lack of camouflage discipline and by the lack of appreciation of the necessity of concealment by men and the lack of instruction and correction to improve their conditions ... The Japanese were outstanding jungle fighters and past masters in the art of camouflage, and we could well emulate their toughness in these things. This gave them an outstanding advantage, which we could overcome by training ... Our mental attitude was that the jungle was a hazard and an obstacle to overcome. To the Japanese the jungle was an aid to his warfare.[43]

General Harding addressed his tenuous situation by first reorganizing his front. The front was broken into two separate forces: Warren and Urbana. The Warren Force occupied the right flank of the front, in the vicinity of the New Strip and Duropa Plantation. It comprised the 1st and 3rd Bns. of the 128th IR and the 1st Bn. of the 126th IR. In support were five Australian guns, three 3.7-inch howitzers and two 25-pounders. The Urbana Force operated on the front's left flank in the vicinity of Buna Mission and Village. This force had the understrength 2nd Battalion of the 128th IR and the 2nd Bn. of the 126th. Two 25-pounders were in support. In the

43 Army Ground Forces, *Observers Reports w/ Appendixes: New Guinea*, 4-5.

Australian Gona sector the 126th IR's 3rd Bn. and whatever divisional support assets in New Guinea were attached. The 127th IR didn't arrive on the front until the second half of December. It is important to note that Harding in these first weeks had only half of the combat strength of his division and few of its support elements. Only five of the nine infantry battalions of the 32nd operated in this first phase of the battle.[44]

Though his Divisional Command Post was undoubtedly getting bombarded with messages from MacArthur, Harding waited until 26 November to strike again. On the Warren Front, Cape Endaiadere and the New Strip were the objectives of a major attack. Eight Australian guns and some heavy 81-mm mortars (heavy due to their extra firepower when compared to the 60-mm weapons used beforehand) in addition to 35 aircraft would support the attack. The 3rd Bn, 128th IR would drive to Cape Endaiadere, the 1st Battalions of the 126th and 128th to follow on the left, hoping to exploit any success. An Australian Company, the 2/6 Independent Company and a company from the 128th IR, would each attack an end of the New Strip. The battle would begin at 0800 on the 26th.[45]

The plan seemed sound. The aircraft bombed and strafed the target area. The guns and mortars followed. The infantry then attacked. The preparatory fires did little. The attack once again met withering machine gun fire and no gains were made.[46] On the Urbana Front the attack fared no better. The men of the 32nd were tired and frustrated. The frustration centered on an undeniable fact. They lacked the weapons they needed.

Samuel Milner, who in preparing the Army's official history interviewed both soldiers and officers that toiled in this campaign, noted:

---

44 32nd Infantry Division, *After Action Report: Buna*, 8.

45 Milner, *Victory in Papua*, 180-2.

46 Military Intelligence Division, *Papuan Campaign: The Buna-Sananda Operation, 16 November 1942 – 23 January 1943*, 33.

> Most frustrating of all, however, was the realization that they did not have the proper weapons to reduce the bunkers that stood in their way. They were without tanks, grenade launchers, or flamethrowers, and mortars, artillery, and air bombardment seemed to have no effect on the enemy's formidable bunker positions. About the only method they could use to reduce the enemy bunkers was to crawl forward as close as the Japanese protective fire would allow and then make a sudden rush in the hope of getting close enough to push hand grenades through the firing slits. This was a course in which, as one who was present was to observe, "Many more failed than succeeded, but for the most part, there was no other way."[47]

The 105-mm. American howitzer, the one American gun made available at Buna, could have made an immediate difference if employed. It's high angle of fire and delayed action fuses would have been effective against the Japanese bunkers. The problem: lack of ammunition. Originally allocated only a few hundred rounds, they were largely gone in the first few attacks. Most of those rounds were fired unobserved, so they were essentially wasted. General Waldron, the Division's Artillery Commander, pleaded for ammunition to I Corps, incensed that his one reliable gun stood dormant for days at a time. The I Corps Ordnance Officer shared Waldron's anger and in a message to him stated: "I've been burning the airwaves since 2 December to have 800 rounds of 105-mm., ammunition flown (to you). The stuff has been at Brisbane airdrome since the night of 3 December ..." [48] Waldron's response: "The General asked for 100 rounds per day for 10 days starting 5 December, and there isn't a single damned round here."[49] Nor were General Harding's requests for

47 Milner, *Victory in Papua*, 197.

48 Ibid., 246.

49 Ibid.

more support granted. He asked for tanks and was given Bren Gun Carriers. He asked for another ten guns, he was told he could have four, "sometime" in December. He asked for his other regiment, the 127th, but was told by General Herring, the Australian commander and his superior, "I cannot see what it is needed for, as you seem to have ample reserves."[50] General Richard Sutherland, MacArthur's Chief of Staff, agreed with Herring, though his opinion centered around the lack of air assets necessary at the time to move the 127th IR to the battle zone. But, he did confer with Harding; he did see the state of his no longer green but still tired and ill-equipped troops. No record exists of Sutherland informing MacArthur of the actual conditions at the front. The calls within MacArthur's headquarters to replace Harding now began in earnest.

FIGURE IX. Harding asked for tanks and was given Bren Gun Carriers. All were easily knocked out.

General Eichelberger noted these calls. After the war he revealed to Milner his animosity about them and along with those calls his anger with General Kenney and his air force. Eichelberger told Milner:

50 Ibid., 200.

> These facts are stated here in my present consideration of his greatness as a General and I consider them in view of the fact that he did not come to Buna although it was vital to him and it was only 40 minutes away by airplane from his headquarters in Port Moresby. Neither did his Air Chief Georg Kenney come over the mountains to Buna 40 minutes away to see whether his bombs were being dropped on the Japanese or on my troops and when I had reported the losses we were taking from American bombs I received neither reaction nor apology.[51]

Despite heavy losses and their growing frustrations, the infantry continued to fight. After the failure of the 26 November attacks General Harding rested the men and prepared to launch a coordinated attack on the 30th. His men rested but the interlude did nothing for ammunition and food shortages.[52] On the Urbana Front, they would move towards Buna Village after a half hour artillery and mortar preparation. Unlike previous attacks they moved into position the previous night and launched their attack before daybreak. Supply shortages were not limited to weapons, ammunition, food, and medicine. There was no white cloth for the men to wear at night so units could follow each other. Instead, these brave men, men that had never received training for night actions, let alone night actions in the jungle, held each other, man to man, as they moved forward. When the attack started chaos was everywhere. Units were out of position, men mired in the swamps. However, Colonel Mott, the Urbana Front commander, and his men pushed forward. Japanese outposts were overrun, and actual progress made. One Company, F Company of the 128th IR's 2nd Bn, secured ground that isolated Japanese forces in Buna from

---

51 General Robert Eichelberger, *Eichelberger to Dr. Samuel Milner, Personal Correspondence Concerning His book, Victory in Papua, 8 March 1954*, 47.

52 Ibid., 189.

the Sananganda front to the west. However, most of the attack still gained little ground. The artillery preparation did nothing and the enemy bunkers, undamaged, again stopped the attack. Casualties continued to mount with no help in the offing. General Sutherland recommended on 30 November that MacArthur relieve Harding. MacArthur called for Eichelberger and ordered him to Buna. He wanted Harding gone and offered a warm message to Eichelberger before his departure:

> Bob, said General MacArthur in a grim voice, I'm putting you in command at Buna. Relieve Harding, I am sending you in. Bob, and I want you to remove all officers who won't fight. Relieve regimental and battalion commanders, if necessary, put sergeants in charge of battalions, and corporals in charge of companies—anyone who will fight. Time is of the essence; the Japs may land reinforcements any night. Bob, I want you to take Buna, or not come back alive. He paused a moment and then, without looking at Byers (my Chief of Staff), pointed a finger. And that goes for your Chief of Staff too. Do you understand?[53]

Eichelberger headed for Buna and when arriving did not immediately relieve Harding. He first assessed the situation himself.

Eichelberger did relieve General Harding. He did not like the condition of the men or that of the commanders on the scene. Eichelberger did acknowledge how battered the 32nd's line units were. He noted that the "126th and the 128th Infantry were badly depleted by illness and wounds, companies with a strength of two hundred and fifty at full muster were down to forty and fifty men, and attrition among line officers was tragic."[54] He still may have kept Harding in command if he indicated he would shake up his

53 Eichelberger, *Our Jungle Road to Tokyo*, 37.

54 Ibid., 55.

command structure. Harding's response: "I'm relieving nobody. The men should not be relieved, they deserve to be decorated."[55] His Chief of Staff, General Meyers, observed Harding's headquarters and saw nothing but defeatism and despair. Eichelberger's old friend left him no choice but to order his relief. Eichelberger, as the I Corps commander, took direct control of the whole front, a force of an understrength division. After observing another fruitless attack on 2 December, he not only relieved Harding but also his regimental commanders. He could see the obstacles to a successful attack. For that Harding and his regimental commanders could not be blamed. They, along with the physical conditions of Buna itself, could be blamed for the deplorable condition of their men. The War Department's Military Intelligence Division's analysis of the campaign noted:

> Even before the attack on 2 December, our troops were tired and dispirited, and this last repulse reduced their confidence to a low ebb, probably the lowest of the entire campaign. Two weeks of fighting had not even dented the Japanese line. Rations had been so short that troops received only one-sixth of a "C" ration per day. Torrential rain alternated with jungle heat. The insects seemed as determined as the enemy. Casualties from disease and wounds had reduced all battalions to approximately half-strength.[56]

General Eichelberger would set 5 December as the date for his first attack. That would give the troops a few days to rest and new commanders some time to settle into their jobs. Some "support" did arrive for the attack: five Bren Gun Carriers instead of the tanks General Harding repeatedly requested, were that support.

---

55 Samuel Milner, *Milner Interview with General Eichelberger*, 21 November 1948.

56 Military Intelligence Division, *Papuan Campaign: The Buna-Sanananda Operation, 16 November 1942 – 23 January 1943*, 35.

These gun carriers could carry guns. They were supply vehicles. Lightly armored, slow, and open topped; they were vulnerable to snipers and ambush and could do nothing against bunkers. Employed on the Warren Front all five were quickly put out of action. The attack there, as usual, failed. Even a flamethrower, a weapon if a soldier got close enough to a bunker could eliminate it, did not work. On the Urbana Front they did manage to cut off Buna Village. That was the extent of the day's success. General Eichelberger now appreciated what he faced. Looking over the battlefield he commented:

> ... The apex of the triangle (the road junction) faces generally south. The area to the west of the triangle is a dense sago swamp. East of it is a grassy plain at least two hundred yards wide. The triangle itself is on comparatively high ground. The terrain at the apex of the triangle is fairly open but the base of it is heavily wooded. The Japanese have set up in this triangle, bunkers made of heavy logs and earth, with connecting fire trenches. I saw some bunkers.. yesterday on our right flank, south of Cape Endaiadere. Some of them could be fired from and some could not but all in all they have presented suitable cover for personnel for anything we have used against them except heavy aerial bombs and direct hits from a 105-mm. howitzer using delayed action fuses. I have tried repeatedly to take these positions without success. The first day I laid down a ten-minute barrage, using 81s and the only 105-mm. howitzer I have, plus a battery of Australian 4.5s, then dumped in a lot of smoke and attacked. No gain was made. We tried the same thing the next day and still no gain was made.[57]

57 Robert Eichelberger, *Eichelberger Diary*, 1942, Dec. 10 Entry.

The above describes why the first three weeks at Buna were such a disaster. Reinforcements and tanks, especially tanks, had to arrive before he attacked again. That would happen around 16 or 17 December. Eichelberger would be patient. During this interlude the 127th IR began arriving, as did veteran Australian infantry and M3 Stuart tanks. General MacArthur had no such patience. In a letter to Eichelberger, dated 13 December, he shows his historic ambivalence to what men he commanded faced:

> GENERAL HEADQUARTERS
> Southwest Pacific Area
> Office of the Commander in Chief
> 13 December 1942
>
> Dear Bob:
>
> Time is fleeting and our dangers increase with its passage. However admirable individual acts of courage may be; however important administrative functions may seem; however splendid and electrical your presence has proven; remember that your mission is to take Buna. All other things are merely subsidiary to this. No alchemy is going to produce this for you; it can only be done in battle and sooner or later this battle must be engaged. Hasten your preparations, and when you are ready—strike, for as I have said, time is working desperately against us.
>
> Cordially
> /s/ MacArthur
> Douglas MacArthur
> Allied Commander[58]

58 Eichelberger, *Our Jungle Road to Tokyo*, 56.

To his credit General Eichelberger did not waver, though he did echo some of the comments of his absent superior. Eichelberger stated: "One of the greatest difficulties we find up here is a shortage of junior officers and non-commissioned officers with guts and ability to lead small units. I would give what little fortune I possess to have a number of good corporals, sergeants, and lieutenants to conduct proper scouting and patrolling here."[59] He did not attack until 18 December. On that date everything was ready.

The tanks would operate primarily on the Warren Front. The area in and around the Duropa Plantation was dry and the heavier Japanese guns were concentrated at Buna Village. The 105-mm. howitzer, with 800 rounds of delayed action ammunition (it finally arrived) would support the attack. After a month of fighting the Americans knew Japanese position locations. The Stuarts, though armed with only a small 37-mm. gun, could destroy the bunkers at close range. Even its thin armor was enough to render the Japanese machine guns ineffective. Veteran Australian infantry would accompany the tanks in the first attack.

The following is an excerpt from a chapter written by this author, *Stalemate to Victory: Combined Arms in World War II's New Guinea Campaign*, as part of the Army University Press' *Historical Case Studies of Combined Arms Maneuvers in Large-Scale Combat Operations*:

> ... The 126th and 128th Infantry Regiments would initially provide support before assuming the brunt of the responsibility. They spent the days since the failed 5 December attack resting, patrolling, and hoping. It would be a multi-pronged attack. One thrust would move through the Duropa Plantation to Cape Endaiadere. The other would drive toward the airstrips and the small bridge connecting them. All the guns, though only the

59 Eichelberger, *Eichelberger Diary*, 1942, Dec. 10 Entry.

lone howitzer mattered, opened fire on the Japanese positions that morning. The howitzer was the only gun with the right trajectory, ammunition, and power to do any damage. Tanks were effective at short range. At the longer ranges artillery operated from 75-mm. and 105-mm. guns lacked the power to damage the reinforced bunkers used by the Japanese. The Japanese must have thought another typical American attack just began. Few of their positions were effected. As they looked through their bunker vision slits instead of seeing unsupported infantry launching another frontal attack, tanks entered the battle.

The tanks had an immediate effect. One infantry battalion commander noted: The tanks really did that job. They apparently completely demoralized the Japs ... who fought like cornered rats when they were forced into the open because of having their fires masked when the tanks broke through their final protective line ... There were few holes knocked in the bunker except where the tanks stood off and blasted them at short range with their 37-mm. guns.

The heavy small arms fire, which had stalled the American infantry for a month, had no effect on the tanks. American and Australian infantry, operating with the tanks, eliminated Japanese troops fleeing their bunkers. Two Stuarts were lost, one to a Molotov Cocktail, the other to mechanical failure. The attack continued to within 500 yards of Cape Endaiadere, destroying Japanese strongpoints along the way. The tanks and infantry pivoted towards the Bridge and airstrips.

> The attack against the Bridge and the New Strip was initially less successful. Twenty Japanese bunkers thwarted the attack until the four remaining tanks arrived. Joined by two Australian infantry companies and the 1st Battalion of the 128th Infantry, they overran the most formidable Japanese defensive system on the Buna battlefield. The infantry finished what was not destroyed by the tanks, able to now advance within a few feet of the bunkers since the tanks drew most of the Japanese fire. The solitary howitzer also destroyed many enemy positions now that their exact locations were pinpointed and the proper ammunition available. Some mopping up remained but on the Warren Front, the 18 December attack won the Battle of Buna.[60]

On the Urbana Front the fight would continue for several more weeks. Though Buna Village was taken, the Mission was not. This ground, unlike that on the Warren Front, was unsuitable for tanks. Without the tanks fighting on the Urbana Front continued to resemble that which characterized Buna throughout this campaign. Tough close quarter fighting, frustration, and indecisive results. To break this situation the Urbana Front needed courage and innovation. The 32nd had proven, despite the words of men like the absent MacArthur, they were courageous.

Continued attacks in the heat and foliage against positions they could not see and when they could see them, they could do nothing with, could not be defeated solely with courage and determination. Innovation had to broach the difference. The 126th and 128th IRs were tired and depleted having fought at Buna since the campaign began. By mid-December the 127th IR had arrived. Fresh and

60 Peter J. Schifferle, Editor, *Bringing Order to Chaos: Historical Case Studies of Combined Arms Maneuver in Large-Scale Combat Operations,* (Fort Leavenworth, Kansas: Army University Press, 2018), 99-100.

FIGURE X. Stuart tanks and Australian infantry moving into an attack.

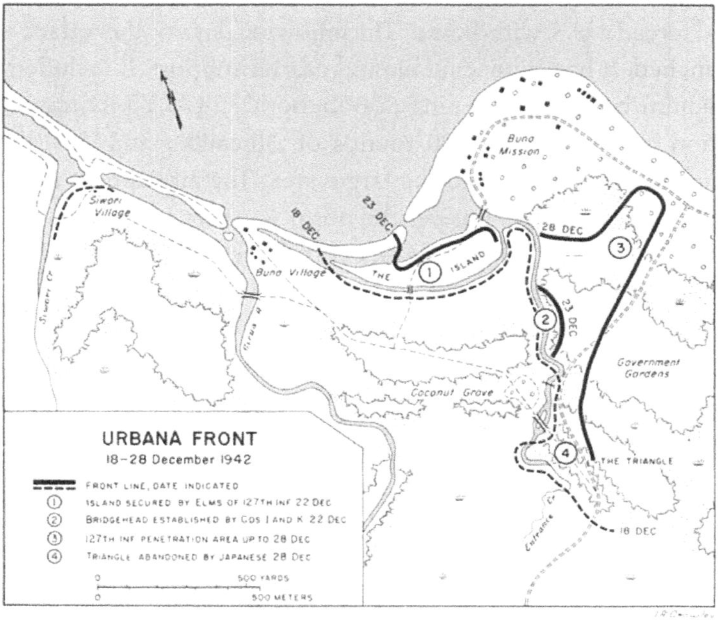

FIGURE XI. The Urbana Front, late December 1942. The Japanese perimeter is being slowly constricted.

at full strength they had numbers but the same problems as their sister regiments. No heavy weapons. The single 105-mm. howitzer was still the only American artillery piece available. The ammunition shortages still existed. Buna Village had fallen but the Japanese occupied Buna Mission and its approaches, including "The Island."

Colonel Clarence Tomlinson commanded the 127th IR. On 17 December he ordered his L Company on the following morning to take "The Island." The problem was getting there. The stream separating it from the 127th was unfordable and the footbridge connecting it long ago destroyed. An audacious plan, to have swimmers drag a heavy cable across the stream, was attempted. It succeeded. Two of L Company's three platoons, plus a machine gun section, were across the stream by noon on the 18th. The Company's third platoon soon followed. All seemed to be going well, too well. As they approached the bridge connecting the Triangle to Buna Mission machine gun fire from hidden positions stopped the advance and forced a U.S. withdrawal. The following day another attack was launched. It had significant air and mortar support. It included 22 medium bombers dropping 500 lb. bombs, 475 20 lb. fragmentation bombs, and 21,000 rounds of .30-caliber and .50-caliber ammunition poured into the target area. The air support certainly looked impressive but per the norm was very inaccurate, doing little damage to the Japanese and coming very close to U.S. forces. Three separate attacks, over one hundred men, were launched and each was easily repulsed. Over 1,000 rounds of mortar ammunition, including white phosphorous shells, did nothing but they had done nothing since the first attack a month earlier. Forty men were killed or wounded. Two more attacks on 20 December by the 127th's E Company also failed. Finally, on 22 December, U.S. engineers had repaired the fallen footbridge between the Island and the 127th's position on the south side of the stream. Men rushed across and by 23 December the Island was secure. Buna Mission remained.[61]

---

61 Milner, *Victory in Papua,* 284-7.

Colonel Tomlinson, after the repeated failures of previous days, requested and was granted his relief. He was replaced by Colonel John E. Grose. On 24 December Grose, with five companies of the 127th IR, was ordered to attack across the Government Gardens towards the Mission. Artillery and mortar support was available though why anyone thought it would work after more than a month of failure is baffling. The infantry advanced. The Japanese were ready:

> Captain Yasuda had the area well prepared for defense and could cover nearly every foot of it with both observed and unobserved fire. The track through the Gardens was covered by bunkers, and on either side of it, echeloned in depth to the rear and hidden by the kunai grass, were numerous individual foxholes and firing pits, most of them with overhead cover. In the surrounding swamp, north and east of the Gardens, were strong bunker positions, and even stronger fortifications were to be found in the plantation and along the shore.[62]

Co. I was the first to be hit as they advanced. Again, despite herculean efforts by individual soldiers, the attack faltered. The 32nd ID earned eleven Medals of Honor during World War II. Two of them were awarded for actions on this day. The first, given posthumously to First Sergeant Elmer J. Burr, stated:

> *... for conspicuous gallantry and intrepidity in action above and beyond the call of duty while serving with Company I, 127th Infantry Regiment, 32d Infantry Division. During an attack near Buna, New Guinea, on 24 December 1942, First Sergeant Burr saw an enemy grenade strike near his company commander. Instantly and with heroic self-sacrifice he threw himself upon it, smothering the explosion with his body.*

62 Ibid., 294.

First Sergeant Burr thus gave his life in saving that of his commander.[63]

Co. I was pulled off the line, replaced by Co. G. They eliminated several Japanese positions but were also stopped by enemy bunkers. As other companies moved through the Government Gardens to the Coconut Plantation, another remarkable act of heroism earned another Division soldier, Sergeant Kenneth E. Gruennert, the second Medal of Honor of the day:

> ***...for conspicuous gallantry and intrepidity in action above and beyond the call of duty while serving with Company L, 127th Infantry Regiment, 32d Infantry Division. On 24 December 1942, near Buna, New Guinea, Sergeant Gruennert was second in command of a platoon with a mission to drive through the enemy lines to the beach 600 yards ahead. Within 150 yards of the objective, the platoon encountered two hostile pillboxes. Sergeant Gruennert advanced alone on the first and put it out of action with hand grenades and rifle fire, killing three of the enemy. Seriously wounded in the shoulder, he bandaged his wound under cover of the pillbox, refusing to withdraw to the aid station and leave his men. He then, with undiminished daring, and under extremely heavy fire, attacked the second pillbox. As he neared it he threw grenades which forced the enemy out where they were easy targets for his platoon. Before the leading elements of his platoon could reach him he was shot by enemy snipers. His inspiring valor cleared the way for his platoon which was the first to attain the beach in this successful effort to split the enemy position.***[64]

63 https://www.32nd-division.org/old/history/moh/32moh.html, Accessed 4 June 2024.

64 https://www.32nd-division.org/old/history/moh/32moh.html, Accessed 4 June 2024.

Elsewhere the attacks accomplished little and the 127th had by the end of the day advanced no more than 150 yards.[65] Col. Grose wanted to reorganize before launching another attack. General Eichelberger ordered him to attack immediately.[66] The tables had turned. Eichelberger urged MacArthur to exhibit patience as he prepared for his major attack. Now he was showing the same impatience as his bombastic senior officer. On Christmas day, Grose, with eight companies, would launch a coordinated attack on enemy bunkers in the "Triangle" and towards Buna Mission. By the end of the day the attacks failed. But the Japanese positions were slowly shrinking and their situation becoming increasingly hopeless.

Several companies were isolated by the attacks of Christmas day and on 26 December. Those companies, along with the entire 127th IR, were wearing down. They now resembled the 126th and 128th IRs after their time in sustained combat. They were tired and frustrated with the lack of progress. A patrol reached those isolated companies and noticed: "The condition of the companies and on our arrival was deplorable. The dead had not been buried. Wounded, bunched together, had been given only a modicum of care, and the troops were demoralized. Major Schroeder did a wonderful job of reorganizing the position and helping the wounded. The dead were covered with earth ..."[67] The attacks continued. By 28 December the isolated companies were reached. The Triangle then fell, and Buna Mission was outflanked. Many bunkers remained and they would be dealt with, either abandoned by the Japanese or eliminated by the 32nd. Essentially, the fighting was done by New Year's Day. The 32nd Infantry Division won the Battle of Buna.

---

65 Milner, *Victory in Papua*, 297.

66 Ibid., 298.

67 Ibid., 300.

CHAPTER III

LESSONS LEARNED, RECOMMENDATIONS MADE

The cost to the 32nd Infantry Division for their Buna victory was high:

Type of Casualty	126TH IR	127TH IR	128TH IR	Service	Totals
Killed in Action	266	182	138	16	602
Died of Wounds	34	29	25	0	88
Died of Other Causes	5	3	4	5	17
Total Deaths	305	214	167	21	707
Wounded in Action:	728	504	1434	14	1,680
Fevers (Malaria)	1,422	2,011	1,264	661	5,358
Common Diarrheas	17	123	203	65	563
Shell Shock	35	89	87	0	211
Misc. Diseases	658	590	682	224	2,154
Total Diseases	2,285	2,813	2,238	950	8,286
Grand Totals	3,318	3,531	2,839	985	10,673[69]

The Division entered the battle with approximately 11,000 effectives. Most support units never made it to the front. The infantry regiments of course took the brunt of the casualties though the support units that did make it to Buna were hit hard by the non-combat elements (950 men out of action at one point or another). The shell shock numbers are drastically low. This ailment was rarely re-

69 I Corps, *History of the Buna Campaign: December 1, 1942 – January 25, 1943*, (Headquarters: I Corps, 1943), 105.

ported nor were any types of psychological problems incurred in sustained combat. In a later chapter when examining the Division on Luzon, when the men were veteran troops, the psychological casualty numbers rose dramatically. Here, in their baptism of fire the numbers are very low. These men were mid-Westerners who had never seen a one-hundred-degree day in their lives. Poorly fed with the constant insects and little sleep, they quickly wore down. They had to learn, as the battle progressed, never a situation a commander wants for green troops; 211 shell shock cases are underreported.

The infantry units saw more than their original number become casualties (most of the non-combat numbers returned to action). Below is the strength of the 126th IR's 2nd Battalion from 2 December 1942 – 5 January 1943:

	Dec. 2, 1942	Dec. 19, 1942	Jan. 5, 1943
	Officers/ Enlisted Men	Officers/ Enlisted Men	Officers/ Enlisted Men
HQ. Co.	4/96	4 /72	2/44
Co. E	4/136	1/40	0/28
Co. F	3/105	1/35	1/31
Co. G	3/127	2/64	1/5
Co. H	4/102	1/61	0/35
Med. Det.	1/17	1/11	1/11
	19/585	10/283	5/154[70]

The above progression is staggering. Already understrength by several hundred men on 2 December the Battalion would see its strength reduced by half every two weeks. The month of December was a month of continuous, sustained fighting. Officers were un-

70 Ibid., S1 Annex.

available in two companies by 5 January. The Battalion's G Company went from 127 to 5 men—5 men! Each of the companies were little more than understrength platoons.[71] Combat was not the only thing reducing numbers. Sickness and fatigue took more. The Division Medical Officer, Colonel Warmenhoven, after the battle finally ended, to determine the condition of the men took the temperatures of 675 infantrymen. He reported that:

> 53 percent of this group of soldiers was running a temperature ranging between 99 degrees to 104.6 degrees. In order of prevalence, the cause of the rise in temperature is due to the following: Malaria, Exhaustive States, Gastro-Enteritis, Dengue Fever, Acute Upper Respiratory Infection, and Typhus (Scrub). The average normal sick-call rate of a command ... was 3.8 percent of its strength. The sick-call rate of the 32d Division was 24 percent and going higher. Some 2,952 men (more than three quarters of them from Buna where the division had made its primary effort) were already hospitalized because of disease and fever, and fifty to one hundred were being evacuated from Buna to Port Moresby daily for the same cause.[72]

Combat losses and the environment whittled down the Division's strength. These men were in the battle zone continuously for seven weeks. No rest, poor rations, insects everywhere. What the Division needed was replacements. None ever reached Buna because at this early point in the war since no trained replacements were available in the United States and the only other American unit in Australia, the 41st Infantry Division, was no more ready for combat than had been the 32nd. General Eichelberger noted that it was not until May 1943 that another U.S. infantry division arrived in Australia. The General gave this answer to a question he was frequent-

71 Ibid., 60.

72 Milner, *Victory in Papua*, 323-4.

ly asked: "Why didn't you relieve those tired men in New Guinea and send in replacements?" His answer: "There were no American infantry replacements in Australia."[73] The 32nd was not ready for Buna, but in war that often does not matter.

Why did MacArthur send the 32nd to Buna? He had to make sure the Japanese did not utilize those airfields, the Old and New Strips. That is beyond a doubt. Another part of his reasoning revolved around the belief this would be an easy fight. Buna was a massive American intelligence failure. The 32nd was led to believe that only a depleted Japanese force remained at Buna. MacArthur's intelligence missed the arrival of thousands of reinforcements and the vast network of bunkers and other defensive positions. As inexperienced and poorly equipped as the 32nd was at this point in the war, when told what awaited them they exuded confidence. That confidence started with the division commander, General Edwin Harding, and went down to the individual soldiers in the infantry squads. Missing the veteran, well-equipped Japanese soldiers in their bunkers cast the 32nd into the nightmare they faced.

Had the Division been better trained and equipped, though an arduous task still faced them, it would have produced a less costly victory. Supplying the men was a nightmare:

> Supply of rations and equipment was an extraordinarily difficult problem throughout the operation. Land transportation across the mountains was almost impossible and very impractical, since there was only a rough and steep foot trail from Port Moresby to the front and a trip over this trail took 18 to 28 days. The distance could be flown by plane in 35 minutes. While units near the coast relied on supply by small boats, those inland had to depend on supply by aircraft.[74]

73 Eichelberger, *Our Jungle Road to Tokyo,* 35-6.

74 Military Intelligence Division, *Papuan Campaign: The Buna-Sanananda Oper-*

I have repeatedly pointed out that only one piece of American artillery, a 105-mm. howitzer, was available throughout the battle. Supplying it with the right ammunition also developed into a problem. A battery of six guns would have undoubtedly helped. How to get the guns forward was the dilemma. They could be flown to the area but moving 6 guns to the front required transport that did not exist. There was not enough ammunition for multiple guns. Again, the intelligence failures affected the operation. If stronger opposition was anticipated better planning, while delaying the operation for a few weeks, would have allowed more support to be in place, as well as more valuable training time given to the inexperienced infantry. MacArthur's intelligence staff failed him but so did his air commander, General Kenney. He convinced MacArthur his aircraft could do all the artillery traditionally did. He even referred to his planes as the "artillery of the jungle." Kenney was wrong.

The troops on the ground were often within feet of Japanese positions. Yet, they did not know how close they were until fired upon. The enemy camouflage and the thick jungle foliage gave the Japanese an enormous advantage. How would a plane spot such a position from the air? The answer is clear: they would not. Positions could not be marked because there were no smoke canisters and even if marked the inaccuracy of bombing demonstrated in this campaign leaves the neutral observer wondering if a marked position would be hit. Other than General Kenney, who like MacArthur, never visited the front, none saw the air support promised before any attack as much of a help. War Department Observers noted, "air reconnaissance was of little use in jungle warfare."[75] Samuel Milner asked General Waldron, the Division artillery commander, how effective was air support? He referred to the 5 December attack, Eichelberger's first as commander, when answering:

---

ation, 16 November 1942 – 23 January 1943, 20.

75 Army Ground Forces, *Observers Reports w/ Appendixes: New Guinea*, Appendix 11, 4.

> On 5 December, most of the bombing was done around the Buna Mission area, and not so much of it around the Village. One of the troubles on the Urbana Front on that day was the fact that the front lines were so mixed that it was dangerous to do very much bombing. We couldn't fix a definite bombing line and do much effective bombing in front of it. Earlier in the campaign, a whole platoon had been practically wiped out because bombing was falling too short, and we did not want to repeat a mistake like that.[76]

General Eichlberger went further: "The Air Force was a great disappointment during the campaign. They were unable at first to locate the Japanese bunkers and did them little damage when they finally got the range. They did however bomb our own troops several times, with the result that they were asked as a favor to take themselves and their support elsewhere."[77] Inaccurate bombing happened in all theaters of the war. Buna was not the exception but the rule.

Colonel Harry Knight was in the Buna area for several weeks both before and during the battle. He commented on many things but his observations on air support are very telling:

> Close support of ground troops by aircraft was a dismal failure in this battle. The dense foliage made the identification of our front lines difficult, and also made it difficult to identify targets for our supporting aircraft. The Air corps now freely admits that close support of ground troops can only be furnished by dive bombers, and that in the

---

76 Samuel Milner, *Milner Interview w/ General Waldron (Artillery Commander)*, Undated.

77 Samuel Milner, *Milner Interview with General Eichelberger*, 21 November 1948, 3.

> jungle, it is necessary that some method of indicating targets and front lines be developed.[78]

This failure was even more pronounced because of the lack of effective heavy weapons on the ground. He further explained:

> In this battle, because the enemy was in a highly organized position, with overhead cover, and because sufficient artillery and mortars were not available on the battlefield, it was especially desirable that front line troops be able to have the Air Corps bomb enemy strongpoints, which were holding up the front-line troops. No joint training had been conducted to teach means and methods necessary to accomplish this close support, and after one disastrous attempt, close support of ground troops by aircraft was not again attempted in this battle.[79]

The 32nd needed support but was content to continue their frustrating fight without the help of the air force because they stood as good a chance of being hit, as did the Japanese.

The problems with air support would be corrected as the war moved forward. In the jungle it never became a viable weapon. Traditional artillery, contrary to the opinion of General Kenney, was the artillery of the jungle. What was not corrected as the war progressed were the mistakes, both in public pronouncements and tactical military decisions, of General MacArthur. Lieutenant Colonel Laurence M. Jones, while attending the Air War College in 1989, wrote *Defense of the Philippines To The Battle Of Buna: A Critical Analysis of General Douglas MacArthur*. It is specific and in this author's humble opinion, a very accurate critique of "America's Caesar." He points out:

---

78 Colonel Harry Knight, Observer Report: Southwest Pacific Theater of Operations, 4 February 1943, (Headquarters: Army Ground Forces, 1943), 6.

79 Ibid., 7.

> Because it was the first land victory of Allied forces over the Imperial Japanese Army, the Buna-Gona-Sananda campaign was of landmark importance to the Allied cause. It was also an opportunity for MacArthur the consummate publicist. He stunned his victorious troops by announcing, "the utmost care was taken for the conservation of our forces with the result that no campaign in history ... produced such complete and decisive results with so low an expenditure of lives and resources." This is an absurdity and an outright lie. The fighting at Buna, Gona, and Sananda cost the Allies 3,300 killed (3/4 Australian) and 5,500 wounded as compared to the six-month Guadalcanal campaign in which American ground forces suffered 1,100 killed and 4,350 wounded. According to D. Clayton James, "the deepest resentment felt by the veterans of the Papuan Campaign was probably reserved for MacArthur's audacity in depicting the casualty rate as relatively light." It should also be noted that during MacArthur's stay in New Guinea between 6 November 1942 and 9 January 1943 he never once went to the front. In fact he never left Port Moresby.[80]

Thus was born the name "Dugout Doug." Even William Manchester, the author of *American Caesar: Douglas MacArthur 1880–1964*, the most acclaimed biography of him, concluded his complete absence from the front was "baffling."[81] Further, Jones points out MacArthur's penchant for criticizing and ignoring the plight of those under him while blatantly lying about his own role and

---

80 *Lieutenant Colonel Laurence M. Jones, Defense of the Philippines to the Battle of Buna: A Critical Analysis of General Douglas MacArthur,* (Maxwell Air Force Base, Alabama: Air War College, 1989), 45-6.

81 William Manchester, *American Caesar: Douglas MacArthur 1880–1964,* (New York: Little, Brown, and Company, 1978), 327.

achievements happened throughout the first year of the war:

> ... probably the most serious of General MacArthur's failings, he never got out from behind the desk to find out what was going on. On more than one occasion this failing was a life-or-death error for the troops at the front. Had MacArthur bothered to visit the Philippine Army in the field prior to the war he would have known they could not oppose the Japanese on the beaches. Had he regularly visited his troops on Bataan he would have realized that his troops were literally starving to death. Had he gone to the front in New Guinea, he would have seen the horrible combat conditions under which he directed his units forward "regardless of cost."[82]

Harsh but true. After the war General Eichelberger also became a harsh critic, though his criticisms tended to remain private.

Eichelberger served MacArthur faithfully both during and after World War II. His Eighth Army was the occupying force in Japan. Yet when speaking with Samuel Milner during his preparation of the official history of the Buna Campaign, he vented as he had never done so before. What burned most in his memory was MacArthur's statement after Buna that "losses were small because there was no hurry. It was one of the great surprises of my life. As you know, our Allied losses were heavy and as commander in the field, I had been told many times of the necessity for speed."[83] He went on: "This I consider about the worst thing MacA ever did to me. To force me to push recruits up against Japanese veterans on a 'Hurry, Hurry' basis with its consequent heavy loss of life and then say that the 'loss of life was small because there was no hurry' is

82 Ibid., 352.

83 General Robert Eichelberger, *Eichelberger to Dr. Samuel Milner: Personal Correspondence Concerning His book, Victory in Papua,* (3 March 1954), 4.

beyond the pale."[84] He summed it up stating: "As I pointed out in 'Jungle Road,' I gained a quick victory in Sananada and it was unfortunate that he did not wait a while because then he could have announced a great victory. This was the beginning of his public pronouncements of 'mopping up' and 'secured.' It became his custom to announce campaigns over before, in some cases, the fight had hardly started."[85] As frustrating as the aforementioned was to Eichelberger he also felt the need to defend the 32nd Infantry Division from the verbal attacks of MacArthur and his chief subordinates: "A great deal has been said and whispered about the 32nd Division, and much of it makes no sense. The 32nd, which 'failed' at Buna, was the same 32nd that won the victory there. No one else did. Later, rejuvenated and retrained, the division went on to establish a superior combat record in the Philippines campaign."[86] The gallantry of this Division can never be overstated. These green, poorly equipped mid-Westerners fought in an unforgiving jungle environment against an experienced enemy in formidable positions. They won without all the organic elements of the Division and without effective air support. Their baptism of fire made them veteran soldiers. When next engaging the Japanese a year later they were battle ready.

84 Ibid., 17.

85 Ibid., 21-2.

86 Eichelberger, *Our Jungle Road to Tokyo*, 38.

CHAPTER IV

REST, REFIT, AND RETRAIN

The 32nd returned to Australia from Buna early in the spring of 1943. They were tired, battered, and depleted. Over the course of the next year, they would receive an influx of replacements in men and equipment. At no point through the end of the war did the Division ever lack the heavy weapons and supplies so rare at Buna. Unfortunately, shortages in numbers plagued them at times moving forward.

Once in Australia the Division set to work addressing the tactical problems so evident at Buna. From patrolling to working with the other arms to leadership training, the Division would be very busy. John Shortal, Eichelberger's principal biographer, published an article titled *Hollandia: A Training Victory*. Hollandia saw the first landings in New Guinea when MacArthur launched his offensive in April 1944. He noted the keys of the training program instituted at Eichelberger's behest:

1. Using the Division veterans to teach the tactical principles of jungle warfare.
2. Chain of command would be actively involved in all facets of the unit from training to combat. Buna showed "inefficiency" in inspecting, supervising, and enforcing discipline as one of the most glaring causes of failure.
3. Conditioning troops for the environmental and situational factors they would encounter in battle. [Eichelberger] realized that all future operations would have to be conducted in the jungle and that the U.S. soldier was not naturally prepared for this environment. Anticipating numerous streams, swamps, and rivers, he insisted that the soldiers learn how to swim and that all units conduct extensive training in stream-crossing operations. Furthermore, since the jungle would be extre-

> mely physically demanding Eichelberger instituted a physical hardening program, which included forced marches, hand-to-hand combat, bayonet training and obstacle courses.[87]

Eichelberger, the veteran commander, knew the process. You build the soldier, both mentally and physically. Then, you teach the needed skills. Finally, you merge the physically fit, mentally strong, well-trained soldiers and their commanders into a cohesive unit. Despite its victory at Buna the 32nd needed to be rebuilt. Eichelberger noted on 9 January 1943, at the end of the Buna battle, that the Division had to be essentially recreated from scratch. In a letter to General Richard Sutherland, MacArthur's Chief of Staff, he stated that: "You are dead right in returning the 32nd Division to the mainland for it will need building up from the bottom ... but when men like Major Smith and Captain Hantelman return to duty there will be something solid to build on and these men have had their baptism of fire."[88] The veterans were the key. The key tactical skills and principles to be taught were night scouting and patrolling, coordination in movement with artillery support, tank-infantry cooperation, and coordination with air assets. Before getting into these specific skills the men arriving in Australia were trained in the United States. That training was based on a series of universal training principles that did not consider the differences between fighting in the ETO and PTO. The tool used to train these new recruits was War Department manuals.

The following words from General George Marshall, the Army Chief of Staff and architect of the American war effort, show his thoughts on training:

> Training means long hours of arduous work. For the officers and non-commissioned officers, it means not only hard physical work but also inten-

87 John Shortal, "Hollandia: A Training Victory," *Military Review*, May 1986, 40-47.

88 Robert Eichelberger, *Eichelberger Letter to Sutherland*, January 1943.

> sive daily study of the manuals covering the latest techniques in warfare. It is only through discomfort and fatigue that progress can be made toward the triumph of mind and muscles over the softness of life to which he had become accustomed.[89]

Most of the American soldiers that fought in WWII were not professionals. They answered the call in time of crisis as Americans always do. The "softness of life" General Marshall refers to is of course the opposite of fighting in the jungles of New Guinea. It takes time and repetitiveness to create the necessary soldier from an American civilian. That is of course one of the main reasons non-battle casualties were such a problem at Buna. For the leaders of the small units, the squads, platoons, and companies that did the bulk of the fighting, Lt. Colonel Joseph I. Greene, in 1942 the Editor of the *Infantry Journal*, commented that:

> The leader of fighting men must never lose sight of the continued training it takes to perfect his troops as individual fighters. He receives them well trained from the training center and then it's his job to perfect them. He must constantly see, during all his units training, that his troops form the fighting man's crafty habits. He must give them hours for special training in these things as they need it. He must take quick steps of criticism and even punishment to rid them of careless, over-confident, and unrealistic habits in training that would invite unnecessary death in battle. He must see that his unit grows into a fighting team made of soldiers.[90]

Col. Greene's words were vital in reconstructing the 32nd ID. It

89 General George Marshall, *Comments on Combat Readiness*, (Washington D.C. Office of the Chief of Staff, 1944).

90 Lt. Colonel Joseph I. Greene, Editor, *Infantry Journal*, 1942.

would receive a major influx of basic trainees and junior officers that had to be integrated into the unit. The training year of 1943 allowed the veterans of Buna to "perfect" the young men arriving from America's training depots. Those training depots taught according to the books, various training manuals that ensured those heading to war zones possessed the same basic sets of skills. That was fine if all these new soldiers went to Europe. Training was very European battlefield centric. What the 32nd experienced at Buna still received but scant attention in War Department manuals. Their real training began when they arrived in Australia.

The Army lives by its manuals. During World War II and to this very day, the basic field manual is *FM 100-5, Field Service Regulations (Operations)*. It covers just about everything though some situations and scenarios are better explained than others. The SWPA was a jungle war and operating in a jungle received only a few pages attention. A specific jungle warfare manual did exist, but it was created for troops operating in the Panama Canal Zone. What little emphasis jungle warfare did get in FM-100-5 is found in its Special Operations chapter. The "special operation" covered the assault of a fortified locality, what became the main task of ground troops in the PTO. To accomplish this mission, it states:

1. A fortified locality may comprise a single, strongly organized position. It will probably consist of a series of strongly organized positions disposed in great depth and breadth in such manner as to be mutually supporting.
2. Reconnaissance involves the employment of reconnaissance and combat aviation, and mobile ground recon units containing engineers, sound, and flash ranging, and signal intelligence units.[91]

The Japanese rarely used mutually supporting positions. They created or selected a point and fought to the death. Attacking such a point is completely different than a defense in depth. It can be

91 FM 100-5, *Field Service Regulations (Operations)*, 1942, 215-217.

both easier and more difficult. Easier because there is no worry about attacks from other positions; harder because the job was not done until the Japanese were eliminated. They did not surrender. They did not care about being outflanked or having a position isolated. The notes on reconnaissance were completely useless in a Pacific jungle. Combat aviation could see little if anything unless flying at tree top level. Radios often could not penetrate the jungle canopy. Infantry squads did reconnaissance. Again, the year of training in theater was invaluable. Imagine a new platoon leader. A Second Lieutenant fresh out of officer's training is now in charge of 36 men. He knows the basics of his organic weapons. His knowledge of tactical principles is based on what he read in FM 100-5 and perhaps a dozen training exercises at Fort Benning's Infantry School. On an early May morning in 1943 his platoon is assigned an attack on a fortified bunker. The bunker is approximately one mile away, its exact location unknown. The entire exercise is in the jungle. The young Lieutenant begins planning his operation. Everything is done in daylight; he has received NO training in night operations. His lessons on conducting reconnaissance are useless. Mobile elements are unavailable; they cannot penetrate the jungle. Air assets are unavailable because they cannot see anything in the jungle. Everything he has been taught seems to mean nothing. Fortunately, the 32nd interspersed veterans in all units and he can rely on their experience. They start patrolling into the jungle. Several hours later they find their target. In a training environment without opposition, they reach their target. Once in the actual attack the manual did help.

FM 100-5 did do an excellent job explaining how to train, organize, and equip a generic assault team. These teams would do most of the work on Japanese fortifications. The manual states:

> The attack echelon is made up of specifically selected troops and weapons. Normally, this attacking echelon will contain infantry, engineers, high muzzle velocity, flat-trajectory guns, and chemical

troops. The engineers equipped with demolition equipment will destroy obstacles and minefields and will assist the forward movement of the infantry and direct fire cannon.

Training of assault detachments is carried out by having them rehearse the contemplated operation on terrain and against fortifications like those to be encountered. The size and number of assault detachments needed in the preliminary operations depend on the size and number of emplacements, which must be reduced. Each tactical grouping must have enough trained assault detachments with sufficient special equipment to insure the reduction of all emplacements, in its zone of action ... Sufficient artillery of all calibers required to reduce the outpost system supports the attack echelon. Artillery of the heavier types assists the preliminary operations by constant bombardment of the hostile main position, paying particular attention to hostile artillery which can bring fire upon troops engaged in preliminary operations ... Some emplacements will remain intact following the employment of heavy artillery and combat aviation. Members of the assault team are assigned to destroy these emplacements by the placing of demolition charges. Under cover of supporting fires and smoke, the assault detachment advances close to the emplacements; final selection is made of the exact location and route for each demolition to be placed; and the preparation of each demolition is completed.[92]

92 Ibid., 218-221.

This is a good framework. Once in Australia and immersed in a vigorous training program, individual units within the 32nd, and all other infantry divisions assigned to the SWPA, could test and tinker with the basic format to make their approach more conducive to the enemy they faced and the environment they fought in.

The main difference between what is described above and the SWPA was the jungle. The jungle requires different tools and different training. *FM 100-5* states:

> The ability to maneuver off trails requires hardened troops, thoroughly acclimated and equipped for close fighting. Their training must enable them to move across country, at least for tactical movements and give them confidence in their jungle technique. Arms and equipment must be designed for maximum mobility through tangled wilderness under oppressive climate conditions.
>
> The rifle and bayonet, automatic rifle, grenade, submachine gun, carbine, machete, and mortars are weapons suited for jungle fighting. Heavy artillery and rockets may replace artillery in the jungle, when lack of observation and the weight of field pieces limit their employment. Against well-constructed defensive positions, flamethrowers are extremely valuable. All infantry heavy weapons must be transported by pack, small cart, or on the backs of men. Tanks may be used in close support of infantry.[93]

Perhaps if *FM 100-5* had outlined an attack on a fortified position in a jungle environment it would have had more immediate effect. No one had experience in such conditions. Later versions of *100-5,* the 1945 edition, addressed this shortcoming. A lack of infor-

93 Ibid., 281.

mation also existed on the interactions of tanks and infantry in a jungle environment.

In 1942, *FM 17-32, The Tank Company Light and Medium,* was published. It appeared before the Battle of Buna and would have been of little use regardless. The tanks that broke the line of bunkers were Australian, as were the infantry that supported them. Three paragraphs mention the use of tanks in the jungle.[94] Those paragraphs were very general though that should not come as a surprise since none had yet to use a tank in a jungle environment. In March 1944, *FM 17-36, Employment of Tanks with Infantry* appeared. The new manual covered in far more detail jungle operations than did *17-32*. Its section on jungle operations explained the conditions necessary for employment of tanks:

1. Against a located enemy resistance which has been holding up the infantry advance, the importance of the resistance being such that its reduction will materially affect the operations as a whole.
2. In sufficient numbers to ensure continuity of effort and to make full use of the tank's own powers of fire and maneuver.
3. After a thorough tank employment reconnaissance has been made.
4. After detailed planning for coordinated action of the combined force participating in the attack. This planning is conducted in echelons down to include infantry elements and tank crews, which are to operate closely together.
5. In conjunction with the supporting fires of all available weapons before and during the attack.[95]

The particulars of the "Attack Against Fortifications" read as follows:

94 FM 17-32, *The Tank Company Light And Medium*, 1942, 14-15.

95 FM 17-36, *Employment of Tanks with Infantry*, March 1944, 77.

> ... medium tanks may be used to assist in the reduction of fortifications ... they assist the infantry by fire and as a protection for the advance ... Precede the attack with heavy artillery preparation ... When paths have been blasted in the obstacles, move the remainder of the platoon of tanks forward by the assault detachment. Tanks fire on pillbox at close range. Tanks pass through the gap followed by assault teams. Assault detachments complete neutralization of pillboxes. Tanks fire at other pillboxes at close range as soon as smoke clears. Assault teams complete capture or destruction.[96]

The use of tanks in this environment is echoed in the October 1944 publication of *FM 72-20, Jungle Warfare*:

> Tanks may be used to advantage against located, limited objectives where the terrain permits. Thorough tank employment reconnaissance is necessary ... Frequently, tanks are attached to infantry companies: the two elements operate together, the infantry providing close-in protection to the tanks. The principal value of the tanks will be in the use of their cannon, flame-throwers, machine guns, and crushing weight in the destruction of enemy bunkers and other field fortifications.[97]

The above notations display an understanding of operating in a jungle environment, albeit for tanks. Rarely was anything larger than a platoon (3-5 vehicles) of tanks working together in any single engagement. *FM 72-20* notes the limitations of terrain. Armor was not always available in the jungle. They also stress the necessity of training with the infantry they supported. The enemy was not Japanese tanks but fortifications.

96 Ibid., 93-4.

97 FM 72-20, *Jungle Warfare,* October 1944, 66.

Artillery was the key infantry support weapon. At Buna, the 32nd had one effective gun. In the future they would have a minimum of four battalions (36 105-mm. howitzers and 12 155-mm. howitzers). Additional guns were added when in active operations, *72-20* also accurately describes the employment and types of guns needed in the jungle:

> ... principles in open areas are equally applicable in the jungle; however, the employment itself presents many difficulties. The value of artillery support is so great, however, that every effort must be made to make this firepower available and effective. All calibers of artillery are desirable. The thick vegetation restricts the effective radius of shell burst so that in general, more large caliber shells are necessary than for similar targets in more open terrain. Artillery should be of sufficient caliber (105-mm. and 155-mm.) to blast away jungle undergrowth and to destroy hostile positions. All weapons should be capable of high-angle fire.[98]

These words mirror the problems faced at Buna. Someone noticed the Australian guns were low trajectory and of little use. One American gun did more damage than two dozen pieces of Australian artillery.

The manuals outlined what was seen as the necessary skills for the soldier at war. With those manuals as a foundation, upon arrival in Australia the 32nd went to work on the jungle centric skills needed. Night scouting and patrolling, especially night scouting, were skills alien to most when arriving in a war zone. Eichelberger, in the same previously mentioned letter to General Sutherland, recalled being dumbfounded by the Division's complete lack of knowledge operating in any way at night. He noted that a soldier told him in

---

98 Ibid., 62.

"20 months of service he had never had one night problem."[99] This was an essential skill since the Japanese did all possible to create havoc around American perimeters once the Sun went down. American artillery and machine guns were far less effective as the fire was inaccurate and unobservable at night. To those completely foreign to soldiering at night in the jungle along with all the insects, snakes, etc., as well as the creepy, constant noises is terrifying. One of the current Army mottos is "we train as we fight." Training at night throughout 1943 eliminated a weakness and at least made the American soldier comfortable in the nighttime jungle.

At Buna, artillery support was scant at best. The lack of numbers was not the only problem. Inexperienced troops, ones that received as little tactical training as did the men of the 32nd, did not know to follow their own fires to give the enemy no time to adjust and prepare for the inevitable attack. Whenever the attack did come the Japanese were ready. The I Corps After Action Report, under "Lessons Learned," noted that:

> Infantry must have had experience of their own artillery firing over their heads and as close in as safety regulations permit. Artillery must have the experience of firing close to their own infantry, to be impressed with the accuracy necessary ... Infantry must have impressed on them the necessity of attacking immediately after the artillery preparation lifts. There were many instances where all effect of the artillery fire was wasted because the enemy were allowed ample time to reorganize and recuperate from the shock of the artillery fire.[100]

As with night actions, working in conjunction with their artillery support was corrected during the 1943 training year.

A few antiquated Stuart light tanks made all the difference at Buna.

---

99 Eichelberger, *Eichelberger Letter to Sutherland,* January 1943.

100 I Corps, *History of the Buna Campaign: December 1, 1942 – January 25, 1943,* "Lessons Learned."

Those were American tanks crewed by Australian tankers with Australian infantry in support. The Japanese, through the remainder of the Pacific ground war, relied on a static defense with prepared positions enhanced by natural terrain. Eliminating those positions was among the most emphasized tasks as divisions like the 32nd prepared to take the offensive. The tanks were now the far heavier, in both weight and armament, M4 Shermans. The tanks and infantry trained, when possible, together. Each became aware of the other's strengths and weaknesses. Japanese armor was rare in the PTO and the tanks the Japanese did possess were small, lightly armed, and prone to various malfunctions.

FIGURE XII. The M4A2 Sherman tank. Weighing 30 tons with a 75-mm. or 76-mm. main gun, 3 machine guns, and 3 inches of armor, it was far superior to the Japanese tanks, of both the light and medium variety, which possessed less than an inch of armor, weighed less than 15 tons, and had no gun heavier than 57-mm.

U.S. tanks in the PTO were infantry support vehicles. They engaged pillboxes and bunkers. They provided direct fire support to infantry attacks. Their armor also made them mostly impervious to enemy automatic weapons and anti-tank guns. However, they were not invulnerable. Even individual Japanese soldiers lying in ambush could immobilize a tank with mines, explosive charges, or something as crude as a Molotov Cocktail. Tanks needed infantry protection as much as the infantry needed their fire support. Further impacting this dynamic was the terrain. The jungle limited, as Buna demonstrated, where tanks could be employed. The infantry then had to deal with the Japanese on their own.

World War II is lauded as the conflict that launched airpower as the decisive weapon of war. There are historians, including myself, that view air power's impact as highly overrated. Strategic bombing is beyond the scope of this examination. Tactical air power is not. In the ETO tactical air power helped the Germans conquer Europe, the British win in North Africa, and the Allies advance from the beaches of France to the heart of Germany. Yet in the SWPA, so vividly displayed at Buna, it was of little use. The aircraft in both the ETO and PTO were excellent, the pilots as the war progressed well trained and experienced. The major difference: nature. Targets such as tanks and concentrations of troops, vehicles, and supplies, even when well camouflaged, became vulnerable on the open ground of Western Europe. War in the ETO was often a war of movement and vehicles on the move are vulnerable to tactical air power. In the jungles of New Guinea and later the Philippines there was no war of movement, at least on Japan's part. Pinpointing a pillbox, bunker, or cave in the jungle is much harder than a moving tank. Then, hitting that target in the jungle, unless using a dive-bomber with the target marked, is also extremely difficult. Fighting between American and Japanese troops occurred in close proximity; sometimes mere yards separated the two. In the coming chapters we will discuss further actions in New Guinea and the Philippines. America had significant air power available, but the jungle prevented its effective use. The 32nd Infantry Division fought on with the weapons they carried or rode with the rest of the war.

CHAPTER V

THE 32ND RETURNS TO COMBAT

After Buna, the Division needed a commander. When General Harding was relieved General Eichelberger became the *de facto* division commander. The new commander was Major General William H. Gill. Gill, before assuming command, had a long and distinguished military career. Born 7 August 1886, he graduated from the Virginia Military Institute in 1907. He then served in the Virginia National Guard until 1912, when he entered the U.S. Army as a Second Lieutenant. He served in the United States until 1915 before embarking on a two-year stint in the Philippines. Returning to the U.S. as a newly promoted Captain just as the nation entered World War I, he spent the next year in the States before embarking for France in May 1918. Upon arrival he was promoted to Major and in the following months distinguished himself in action, earning a Silver Star. Between the wars he did as most Army officers at the time, attending various schools (including Honor Graduate of the Command and General Staff School in 1925) and shuttling between command, staff, and training duties. In 1936, he became Executive Officer of the 27th Infantry Regiment at Schofield Barracks, Hawaii. He assumed temporary leadership of the Regiment in August 1938 until the new commander arrived. His new boss was Colonel Edwin Harding (an incredible piece of irony).

Now a Lieutenant Colonel, he assumed command as Professor of Military Science for Berkeley's Reserve Officers Training Detachment in the fall of 1938. Two years later, promoted once again, this time to full Colonel, he moved to Fort Jackson, South Carolina as the Executive Officer of the 8th Infantry Division. In October 1941, now a Brigadier General, he took over as commander of the 28th Infantry Division's 55th Infantry Brigade. In May 1942, now a Major General, he took command of the 89th Infantry Divisions and supervised that Division's activation and initial training. He did not accompany them to Europe. Instead, he was sent to Australia and given command of the 32nd Infantry Division. When they arrived back in Australia in March 1943 to rest, refit, and retrain, General William H. Gill was waiting for them.[101]

During the fight for Buna the 32nd operated as part, along with the 41st Infantry Division, of the I Corps. As more men and hence more units flowed into the SWPA, a larger organization was needed. The Sixth U.S. Army became that organization. MacArthur chose Lieutenant General Walter Krueger as its commander. Eichelberger, commander of I Corps, in this author's opinion, earned that assignment. Pulling the Buna Campaign from the depths of despair, he had proven his mettle as a combat commander. This is in no way meant to disparage General Krueger's stellar career to his assumption of Sixth Army command.

Walter Krueger was born on 26 January 1881 in West Prussia, Germany. Brought to the United States as a young boy he enlisted in the U.S. Army and served in the Spanish-American War in Cuba, as a Private. He then served in the Philippines during the Filipino Insurrection, earning a battlefield commission as a Second Lieutenant in 1901. He served along the Mexican border in 1916 with a regiment of the Pennsylvania National Guard. During World War I, though he saw no action, he was Assistant Chief of Staff then Chief of Staff of the 84th Infantry Division. In October 1918, he became Chief of Staff of the new American Tank Corps. During the inter-war years, he would rise rapidly (in the lean days of those years) from Captain in June 1920 to Lieutenant General in May 1941. He attended all the schools necessary for advancement and was an instructor at the Army War College. His attempt to transfer to the new Army Air Corps in 1927 ended when his flight instructor flunked him. Before that he attended the Naval War College in 1926 and returned to that institution as an instructor for four years (1928–1932).

101 Edward Jaquelin Smith, *Always A Commander – The Reminiscences of General William H. Gill*, (Colorado: Colorado College, 1974).

FIGURE XIII. General William H. Gill

FIGURE XIV. General Walter Krueger

He then moved around quite a bit as the following so attests:

1932–1934: Commander, 6th Infantry Regiment

1934–1936: U.S. War Plans Division, promoted to Colonel

1936–1938: Chief, War Plans Division, promoted to Brigadier General

June 1938: Commander, 16th Infantry Brigade

February 1939: Commander, 2nd Infantry Division, Promoted to Major General

October 1939: Commander, U.S. Army VIII Corps

May 1941: Commander, U.S. Third Army, promoted to Lieutenant General

After assuming command of Third Army General Krueger led them through the Louisiana Maneuvers later in 1941. He would continue to train that unit through most of 1942 until the call came for him to head to the Pacific. It is questionable why MacArthur would want Krueger, whose Third Army had trained for deployment to Europe, in his theater. Krueger had not been there since he was a 20-year-old Second Lieutenant in 1901. Already 60 years of age when the war began he must have been surprised at 62 when MacArthur requested him to assume command of the Sixth Army. Any questions about his physical stamina were unnecessary; he led Sixth Army until the end of the war.[102]

In early 1944, the 32nd moved from Australia to Sansapor, several hundred miles west of Buna, on New Guinea's northern coast. MacArthur's goal was of course his return to the Philippines. To get there he had to secure New Guinea. His forces did not have to control every inch of the jungle covered, mountainous island. They needed key points that had Japanese constructed airfields or land suitable for the quick construction of airfields, suitable ports/

102 https://militaryhallofhonor.com/honoree-record.php?id=268, 15August 1944.

harbors for staging follow-up landings, and knowledge of Japanese dispositions. There were over 100,000 Japanese soldiers and hundreds of aircraft at various locations. In eastern New Guinea was the Japanese Eighteenth Army of 55,000 men, headquartered at Wewak. Approximately 60,000 more were posted west of Hollandia and on several small islands off the coast (Wakde and Biak, both of which had vital airfields). MacArthur could not advance without those airfields. Navy carrier support would be available for only a few days after the first landings in April. The carriers were vulnerable to Japanese air attacks from dozens of possible locations in not only New Guinea but islands in close proximity. The Navy was also leading the charge in its own campaign across the Central Pacific and needed its aircraft carriers. Where to attack first was the prevailing question. MacArthur and many other U.S. commanders had a secret weapon aiding them in such decisions, *ULTRA*.

Briefly mentioned in a previous chapter, *ULTRA* was the top secret program that allowed Allied cryptologists to read enemy radio traffic. The Japanese Eighteenth Army was part of the New Guinea Area Army. Daily status reports, dispositions, and needs were sent over the radio. Individual locations like Hollandia, Aitape, Biak, etc., all sent status reports over the radio. MacArthur and Krueger, whose army Krueger designated "Alamo Force," would conduct the campaign. They knew exact Japanese strength, locations, and capabilities. *ULTRA* was initially not disseminated below the level of army commander, though several corps commanders in the SWPA were given access. As pointed out by historian Edward Drea, the unquestioned expert on this campaign, "... 6th Army pointed out that in the Pacific Theater, corps operated as armies did in the European Theater and that corps in the Pacific could thus receive *ULTRA* intelligence."[103] Key leaders would wonder why they received orders from Krueger that made little military sense. It was *ULTRA*.

103 Edward Drea, *Defending the Driniumor: Covering Force Operations in New Guinea, 1944*, (Fort Leavenworth, Kansas: Combat Studies Institute, 1984), 23.

Hansa Bay, approximately 100 miles west of Saidor, seemed the logical objective. The harbor was excellent. There were airfields. However, it had a substantial garrison and was supported by the main body of the Japanese 18th Army several hundred miles west in Wewak. *ULTRA* told MacArthur, Krueger, and their planning staffs that the Japanese were very weak in the Hollandia area. Hollandia's outstanding harbor could support the massive buildup of ships and material necessary to continue the trek across New Guinea and eventually the Philippines. But, *ULTRA* told SWPA Headquarters that it was lightly defended. Aitape, one hundred miles east of Hollandia and home to several outstanding airfields was another possible target and also lightly defended. *ULTRA* allowed MacArthur to take an enormous risk. Should he land at Hollandia the Japanese 18th Army would be cut off. All support and reinforcements were hundreds of miles further west with Krueger's army between them. The strategic benefits were obvious but a leap forward of this magnitude could be disastrous, as pointed out by Drea in another book, *MacArthur's Ultra: Codebreaking and the War against Japan, 1941–1945*:

> Hollandia carried enormously high risks. An amphibious attack is the most complicated and dangerous of military operations. MacArthur now proposed to conduct two simultaneous amphibious assaults—later amended to three—far behind Japanese lines. It would be the largest Southwest Pacific operation to date, ultimately involving 217 ships moving 80,000 men with supplies 1,000 miles in order to bypass and isolate the Japanese Eighteenth Army.[104]

Aitape became the necessary second objective after the Navy informed MacArthur it could provide needed air cover for only three

104 Edward Drea, *MacArthur's ULTRA: Codebreaking and the War against Japan, 1942–1945*, 105.

days.[105] Aitape had three Japanese airfields, the Tajidi airstrips. Those airfields could support the forces at both Hollandia and Aitape in addition to the operations being planned further west. Further, the 18th Army could be incessantly harassed. The operations began on 21 April and were a resounding success. The 24th infantry Division landed at Hollandia, the 41st Infantry Division at Aitape. Since the 41st was tasked with imminent landings at Wakde and Biak further west, the 32nd Infantry Division returned to the war. They landed at Aitape and relieved the 41st Infantry Division's 163rd Regimental Combat Team. General Gill described his Division's easy landing and mission:

> On May 4, the 32nd Division, less two regimental combat teams, moved from Saidor, where we had first encountered the Japanese, and landed on Blue Beach at Aitape with me in command. I immediately took command of the Persecution Task Force (name given to the Hollandia force), which was the defense task force in the area, having as its mission the protection of the Aitape airstrip. We were also to locate the main Japanese force and to determine from their operations and their movements what their intentions were. Moreover, we were to prevent the 18th Japanese Army from moving to the west. The reorganization of the Persecution Task Force was completed soon after the 32nd arrived ... All the 32nd Division officers in the area were battle-trained, jungle-trained operators.[106]

The last sentence I found of particular note as thinking back to when the Division faced its first battle at Buna. They were ready for this fight.

---

105 Ibid., 143.

106 Edward Jaquelin Smith, *Always A Commander: The Reminiscences of Major General William H. Gill*, 58.

General Gill would remain in command of the Persecution Task Force until the later part of June. Then when other units arrived a corps headquarters, the XI Corps, would be formed under the newly arrived General Charles Hall. Gill had to protect Aitape and its vital airstrips. East of Aitape was his main line of resistance. The 126th IR was the main force here, along with elements of his other infantry regiments, the 127th and 128th. Most of his divisional support elements also occupied that area. Neither Krueger nor Hall wanted enemy units anywhere near those airfields. Therefore, further east along the western bank of the Driniumor River, a covering force was established. By doctrine the:

> Purpose of such a defensive line as to provide time for the main force to prepare itself for combat, to deceive the enemy as to the actual location of the main battle position, to force the enemy to deploy early, and to provide a deeper view of the terrain over which the attacker would advance. If forced to withdraw, the covering force was expected to fight a delaying action against the attacker.[107]

Two missions for one division. In later pages the lack of troops will be obvious. Reinforcements would come but not quick enough. *ULTRA* could tell Japanese strength and dispositions to a point. It could not analyze their resolve and desperation.

55,000 men of the Japanese 18th Army were cut off when the 6th Army landed at Hollandia and Aitape. Resupply, reinforcements, and/or evacuation seemed hopeless. The Japanese Navy did not have the forces in the area to contest their American counterparts and supply ships were easy targets. Again, *ULTRA* reaped benefits. Japanese naval radio traffic was intercepted and merchant ships attacked. The Japanese 18th Army commander, General Hatazo Adachi, faced two choices:

107 Drea, *Defending the Driniumor: Covering Force Operations in New Guinea, 1944*, 23.

1. Stay where he was and slowly see his men succumb to starvation and disease.
2. Take his force west with the ultimate objective of Aitape in what was essentially a suicidal attack.[108]

Adachi informed his superiors he would begin moving west towards the Driniumor River and Aitape at the beginning of May. *ULTRA* gave this information to MacArthur who passed it down to Krueger. The Japanese planned to attack in the first ten days of July. Two months away. During that time Sixth Army would continue moving west while an enemy army moved against them from the east. The 32nd Infantry Division would be on its own, at least initially.

No serious concerns seemed to arise from either MacArthur's or Krueger's Headquarters. *ULTRA* told them of the weakened state of Adachi's army. Robert Ross Smith, the Army's chief historian of the entire New Guinea campaign and author of the *Approach To The Philippines,* was unaware, unless told by a senior commander after the war, of *ULTRA*'s existence. Its existence was not revealed until the 1980s. He did see MacArthur's daily intelligence summaries and from that noted what Japanese forces were moving west. Listed as an army of 55,000 men, the figure is misleading. Less than half that number was actual combat troops and they were not in great physical condition. As Smith explains:

> ... Many of the 50,000 men of Lt. Gen. Hatazo Adachi were service personnel, others belonged to air force ground units, and some were naval troops, which had recently passed to the control of the 18th Army. The 20th Division's three infantry regiments were greatly understrength and probably totaled fewer than 3,000 trained in-

108 As stated by Lieutenant General Yoshihara Kana, the 18th Army Chief of Staff, :the Japanese could die where they were and run out of supplies by September or die advancing ... They advanced." Ibid., 28.

> fantrymen. The entire strength of the division, including about 1,000 artillerymen (without guns) and other organic or attached troops, was about 6,000 at the end of May. The 41st Division contained less than 4,000 infantry effectives and a total strength of some 10,700. The 66th Infantry of the 51st Division, also scheduled to participate in the attack on Aitape, did not number more than 1,000 men. Altogether, General Adachi mustered for service in the attack about 20,000 troops.[109]

Their supply situation was even worse. Those 20,000 men had:

1. 13,142 rifles
2. 726 machine guns
3. 561 grenade launchers
4. 22 light mortars
5. 36 75-mm. mountain guns
6. 42 70-mm. guns

- The artillery had only half its needed ammunition and the guns had to be dragged through the jungle. Few guns made the entire trip.
- Communications equipment was nearly gone.
- There were serious shortages of clothing, blankets, and mosquito nets (led to large cases of malaria).
- There were few medical supplies.
- Their food would be gone in August.
- They had no motorized transport, though even if they did have that transport it would have been useless in the jungle.[110]

109 Robert Ross Smith, *The Approach to the Philippines,* (Washington D.C.: Center of Military History, 1953), 147.

110 Ibid.

With every step the Japanese took west their numbers decreased. The troops that remained saw their physical condition worsen. *ULTRA* was aware of all this. Then all went silent. The Japanese no longer communicated, through radio, to their higher headquarters or even among the 18th Army's subordinate units. That spread some panic at *ALAMO* Headquarters but *PERSECUTION* Task Force Headquarters, responsible to hold the Driniumor River line, was deeply concerned. General Gill's 32nd ID was still responsible for this "Covering Force" and his Division was scattered from the Driniumor back to Aitape. Gill was initially told he could expect a Japanese attack at the end of June but it was pushed back to 10 July.[111] He needed reinforcements.

The 32nd found itself in a precarious position. Again, MacArthur and through him Krueger, entered the equation. Krueger's Sixth Army was MacArthur's main weapon and MacArthur was looking west, not east at an enemy hundreds of miles from his main bases at Hollandia and Aitape. Regardless, 20,000 men heading towards him had to be addressed. There were few forces immediately available to reinforce the 32nd. The 6th Infantry Division was fighting in the jungle west of Hollandia at Sarmi. The 24th Infantry Division was still at Hollandia, preparing to move even further west. The 41st Infantry Division was in a vicious fight on the island of Biak. The *Persecution* Task Force had at the end of 3 June, only 3 infantry battalions (2 from the 128th Infantry Regiment, 1 from the 127th Infantry Regiment) guarding the Driniumor River line from the coast to the small hamlet of Afua some two miles south. Further, a heavy 155-mm. artillery battalion and significant air power was withdrawn to support western New Guinea battles. At the beginning of July, with the Japanese moving closer and closer the withdrawn support returned and was bolstered by more guns and aircraft. What the line needed was infantry reinforcements. Elements of the 43rd Infantry Division could arrive after 11 July, the time of the Japanese attack. Krueger held the 31st Infantry Division back

111 Ibid., 133.

since it was earmarked for a later operation. The only unit immediately available and it arrived the first week of July was the 112th RCT. It would bring its own problems.

While General Gill remained the *PERSECUTION* Task Force Commander, it was his Assistant Division Commander, General Clarence Martin, in immediate command on the river. His line's western flank was anchored on the sea, his eastern flank at Afua, a few miles south. To defend this he had on the river theoretically two regiments. Three of the 32nd's infantry battalions were in the west, the newly arrived 112th RCT at Afua. It was weaker than it sounds. A cavalry regiment was far less powerful than a standard infantry regiment. It had two squadrons (the cavalry equivalent of a battalion) instead of the infantry regiment's three; its squadrons had three troops (the cavalry equivalent of a company) instead of four and only one heavy weapons troop instead of an infantry regiment's three.[112] The heavy weapons company had the heavy machine guns, mortars, bazookas, and anti-tank guns needed for immediate fire support. The entire 112th had approximately 1,500 men as compared to the 3,000 of an infantry regiment. General Martin therefore had five battalions to defend his line. As Drea in *Defending the Driniumor* points out:

> By doctrinal theory, a rifle battalion in the defense would be assigned frontage of 900 to 1,800 meters, depending on the defensive strength of the terrain. The depth of such a defense would vary from 720 to 1260 meters ... (before the arrival of the 112th) the 1st Battalion, 128th Infantry, held a 1,800-meter front running south from the mouth of the Driniumor. The 1st Bn, 127th Infantry, defended a 3,300-meter front that extended north from the vicinity of Afua. Patrols covered the approximately 2,700-meter gap between the units. The two rifle

112 Ibid., 135.

> battalions were greatly overextended and, given the jungle terrain, were in fact isolated.[113]

With 20,000 enemy troops, even poorly supplied and physically debilitated, approaching a line held by less than 1/5 that number, the situation was dangerous. U.S. patrols encountered sporadic Japanese troops but not Adachi's main body. They had gone silent. ULTRA was useless if the enemy did not send out multiple radio messages. A 25th June intercept had the night of 10 July as the time of the attack.[114] It would happen that night though Adachi's written orders to his units were issued on 3 July:

1. The 18th Army will attack the enemy occupying the Driniumor River, drive back the enemy in the Pawp and Afua vicinities, and then advance to the Nigia River to prepare for the next or main attack against the enemy's main positions.
2. The attack will begin at 2200 on 10 July.
3. The 20th and 41st Divisions will attack in column formation, cross the Driniumor, annihilate the American defenders on the covering force, and complete the operation by 12 July.
4. Both divisions will finalize their preparations for the main attack against Aitape by 15 July.[115]

These were written orders. In the days before the attack nothing was broadcast over the radio. The Japanese had also pinpointed the three battalions defending the river. It was supposed to be five but on the expected day of the attack, General Krueger wanted the Japanese found and ordered, through General Hall, a two-battalion reconnaissance in force along the northern and southern flanks of the line. Krueger made the Japanese attack far easier for them.

---

113 Drea, *Defending the Driniumor: Covering Force Operations in New Guinea, 1944*, 24.

114 Ibid., 37.

115 Ibid., 64.

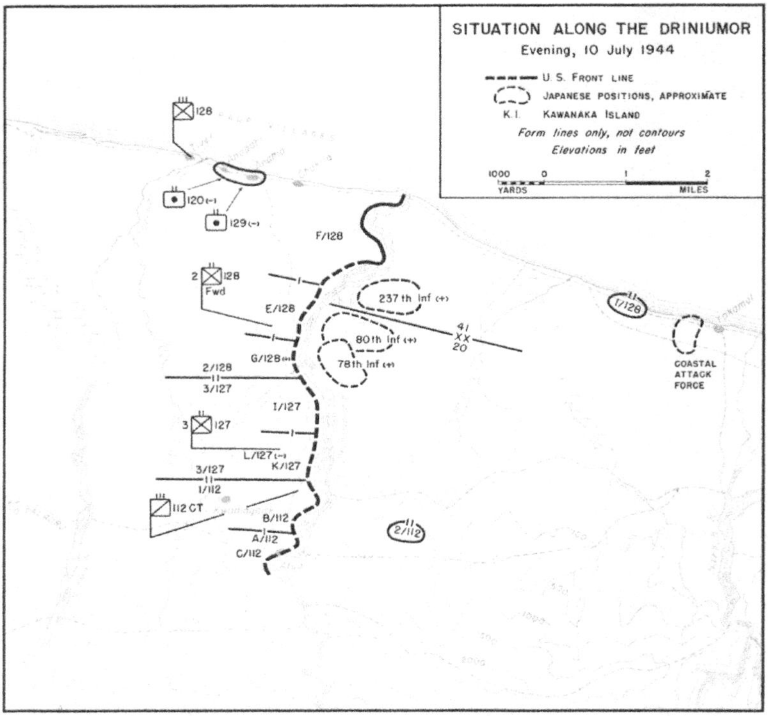

FIGURE XV. U.S. Japanese dispositions along the Driniumor on the eve of the main Japanese attack.

As can be clearly seen above, three U.S. infantry companies would absorb the main Japanese attack. Not a division, not regiments, not even battalions, but companies. At full strength a company had 144 men. 20,000 were about to come screaming across the River.

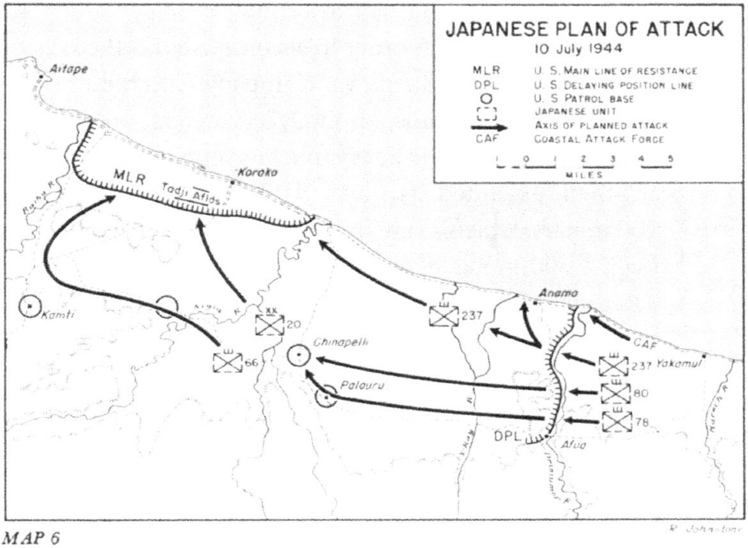

MAP 6

FIGURE XVI. Japanese Plan of Attack

As is clear from the map above the Japanese planned to attack what was the weakest point of the American line, held by only three infantry companies, and then fan out to the left towards Afua and to the north towards the coast.

As must be noticed by now, I rely heavily on Drea and Smith for the Driniumor source material. The records of the 32nd and its subordinate infantry regiments are usually superb. For this campaign they were a bit sparse. Smith's work is in essence a primary source as he conducted extensive interviews after the war with those survivors involved in the battle. Drea covers the broader strategic factors and decisions of higher commands. He also describes the ever-worsening situation for the American units on the Driniumor:

> Until 9 July, the 2nd Bn's sector had about a 1,600-meter frontage, with the 1st Bn, 128th INF, and the 3rd Bn, 127th INF, covering its northern and southern flanks, respectively. When General Krueger ordered the 1st Bn to conduct a reconnais-

> sance in force, 2nd Bn had to occupy the line vacated by the 1st Bn. In order to accomplish this the 2d Bn committed its reserve, Company F, to hold the 1st Bn's former line, about 2,700 meters long. Company E, with one heavy machine gun platoon attached, extended its north flank an additional 900 meters, which made the company responsible for a 1,600-meter defensive frontage, normally a battalion size sector. Such extended frontages and the many small islands covered with dense foliage in the Driniumor made an adequate defense impossible.[116]

The 1st Squadron, 112th Cavalry now occupied nearly two miles of the line. Moving left the 127th's 3rd Bn had another 1.5 miles of front. The 128th's 2nd Bn had over three miles to cover. They had two of the three division artillery battalions (the third was attached to the 41st ID on Biak), a battalion of 155-mm. howitzers, and some tank destroyers. Their machine guns were locked in. Barbed wire was strewn across the front, hoping it would slow the Japanese down. The reconnaissance in force battalions continued into the jungle going right past the Japanese moving up to attack. It was not the infantry's fault. The jungle was so thick no one could see anything. But, at the river that night they heard the Japanese coming.

Except for a few somewhat clear pieces of land, the battlefield here was overgrown with jungle. This was, as usual, a horrible place to fight a battle. It is of no wonder that the two battalions sent east for the Reconnaissance in Force walked right past the Japanese without seeing each other.

Sporadic Japanese artillery fire signaled the start of the attack. But, the men along the river needed no signal. The Japanese moved into the shallow river in great masses of humanity, screaming to announce their presence. American artillery, mortars, heavy and

---

116 Ibid., 69-70.

FIGURE XVII. The Driniumor River

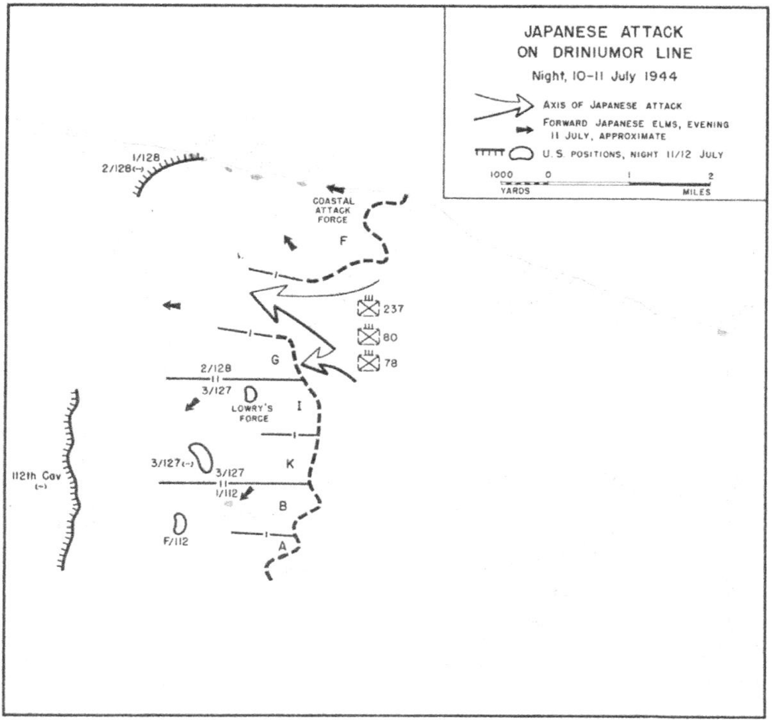

FIGURE XVIII. The Japanese Breakthrough

light machine guns, and standard infantry weapons took a fearsome toll. Firepower was all the men could count on due to their meager numbers all along the line. Companies E and G took the initial brunt of the main Japanese attack. In an amazing story a single American BAR operator (Browning Automatic Rifle) fired over 500 rounds in only 15 minutes into the human wave attacks. His weapon overheated, as did many other weapons that night.[117] Fortunately for the 32nd ID units deployed along the river the Japanese attack was poorly coordinated. Otherwise some 10-20,000 men could have advanced unimpeded to the vital airfields only 35 miles distant.

General Martin had no reserves to throw into the battle. When Companies E and G, at some points outnumbered 10 to 1 and very low on ammunition, had to pull back, large gaps in his line appeared. Martin did receive permission from General Hall to recall the two battalions he had been forced to send out in the ill-fated reconnaissance in force. Martin had no choice but to fall back to his next delaying position, River X. General Krueger was not happy with that decision; he did not feel a withdrawal was necessary. This author cannot comprehend how.

The Japanese did break through on that first night as planned. However, they neither reached Afua to the south nor the coast to the north. Japanese losses were massive, and units quickly became scattered about. Within a few weeks all had been killed, captured, or withdrew back east.

Krueger could question Martin's decision to pull back. He had an enormous hole in his line and had to stabilize his front. In a MacArthur type situation, he was nowhere near this front. He was dealing with operations further west. The American situation greatly improved after that first night. The Japanese suffered heavy losses and those enemy troops that did break through became badly scattered. Further, reinforcements began arriving. The two recon-

117 Robert Ross Smith, *The Approach to the Philippines,* 73.

naissance battalions were threatening the rear of 18th Army and a fresh infantry regiment, the 31st Infantry Division's 124th, arrived at Aitape and quickly moved forward.

Two Medals of Honor were bestowed to the 32nd Infantry Division for this campaign. One was given to Staff Sergeant Gerald Leon Endl, who was in the reconnaissance in force and displayed incredible bravery in an engagement when returning to the Driniumor River. The citation reads as follows:

> ... SSG Endl was at the head of the leading platoon of his company advancing along a jungle trail when enemy troops were encountered and a firefight developed. The enemy attacked in force under heavy rifle, machine gun, and grenade fire. His platoon leader wounded, SSG Endl immediately assumed command and deployed his platoon on a firing line at the fork in the trail toward which the enemy attack was directed. The dense jungle terrain greatly restricted vision and movement, and he endeavored to penetrate down the trail toward an open clearing of Kunai grass. As he advanced he detected the enemy, supported by at least six light and two heavy machine guns, attempting an enveloping maneuver around both flanks ... Twelve members of his platoon were wounded, seven being cut off by the enemy. Realizing that if his platoon were forced further back, these seven men would be hopelessly trapped and at the mercy of a vicious enemy, he resolved to advance at all cost, knowing it meant almost certain death, in an effort to rescue his comrades. In the face of extremely heavy fire he went forward alone and for a period of approximately ten minutes engaged the enemy in a heroic close-range fight, holding them off while his men who were lying along the trail,

> one by one he brought them back to safety. As he was carrying the last man in his arms he was struck by a heavy burst of automatic fire and was killed.[118]

Selfless sacrifice is the norm, not the exception for men in combat. This just happens to be one that was recorded. SSG Endl was an American hero.

For the next several days' American forces would begin eliminating Japanese detachments west of the Driniumor. It was not being done quick enough to suit General Hall, who made his displeasure known to General Gill in the following message. In a previous chapter, I presented the so-called speech MacArthur gave to Eichelberger before sending him to Buna. Hall's message was equally ridiculous, showing another out of touch commander (the entire message was capitalized):

> 16 JULY 44 0005
>
> TO: GILL, CG, PCF
>
> FOR EYES OF GENERAL HALL
>
> CAREFUL ANALYSIS OF YOUR SITUATION DISCLOSES AT YOUR DISPOSAL 127, 128 (-), 124 (-), 112TH, SOME TD'S AND ENGINEERS. CONFRONTING YOU WEST OF DRINIUMOR IS AN UNDETERMINED NUMBER OF JAPS BUT CERTAINLY NOT YOUR EQUAL IN EITHER NUMBER OR FIREPOWER. IT APPEARS TO ME THAT WITH THE FORCES AT YOUR DISPOSAL IF THE PROPER OFFENSIVE ACTION IS INSTITUTED AT ONCE YOU SHOULD BE

118 https://www.32nd-division.org/old/history/moh/32moh.html, Accessed 4 November 2024.

> ABLE TO CLEAR THE AREA WEST OF THE DRINIUMOR WITHIN 48 HOURS. TROOP MOVEMENTS HAVE BEEN DELAYED AND I CANNOT PROMISE YOU ANY REINFORCEMENT. I DO NOT ATTEMPT TO TELL YOU WHAT MEASURES TO TAKE TO COUNTER AN OFFENSIVE FROM THE EAST AND AT THE SAME TIME CLEAN UP THE SITUATION WEST OF THE DRINIUMOR. IT CANNOT BE DONE BY DEFENSIVE ACTION. I EXPECT YOU TO TAKE ALL OFFENSIVE MEASURES NOT ONLY TIO CLARIFY THE SITUATION BUT TO ERADICATE HE ENEMY WEST OF THE DRINIUMOR. YOU MUST DO IT WITH YOUR OWN FORCES WHICH ARE CONSIDERED ADEQUATE FOR THE PURPOSE. WE CANNOT WASTE TIME BY DILATORY TACTICS. WHILE I APPRECIATE THT SOME OF YOUR TROOPS ARE TIRED I KNOW OF NO BATTLE WHICH WAS ENTERED INTO WITH FRESH TROOPS. PLEASE GIVE THIS YOUR PERSONAL ATTENTION AND PUSH IT TO A CONCLUSION IN ORDER THAT INCOMING TROOPS MAY BE USED TO FINISH UP THIS SITUATION OUTSIDE OF YOUR AREA.[119]

It sounds just like MacArthur imploring his Buna forces to attack: "Do it with your own forces, which are considered adequate for the purpose." Considered adequate by whom? "We cannot waste time by dilatory tactics." I can imagine words like the aforementioned infuriating General Gill though he does indicate as such in his memoirs. Hall's last sentence, the one which followed the

119 Drea, *Defending the Driniumor: Covering Force Operations in New Guinea, 1944*, 95-6.

statement he knew some of the 32nd ID's troops were tired, did demonstrate the mindset of the upper levels of command in the SWPA: "Please give this your personal attention and push it to a conclusion in order that incoming troops may be used to finish up this situation outside of your area." SWPA never seemed to focus on the situation at hand. Operations throughout New Guinea, separated by many hundreds of miles, occurred simultaneously. There was never enough men or support for everyone. Regardless, the Japanese 18th Army retreated east and any remnants still west of the Driniumor were finished off by July's end. The 32nd was pulled back; other units finished off Japanese forces to the east.

While fighting in the remaining days of July, clearing Japanese from both sides of the river, another Medal of Honor was awarded to a Division soldier. Private First-Class Donald Ronald Lobaugh of I Company, 127th Infantry Regiment received this highest honor for actions at Afua on 22 July. His citation read:

> ... While Private Lobaugh's company was withdrawing from its position on 21 July, the enemy attacked and cut off approximately one platoon of our troops. The platoon immediately occupied, organized, and defended a position, which it held throughout the night. Early on 22 July, an attempt was made to effect its withdrawal, but during the preparation therefore, the enemy emplaced a machine gun, protected by the fire of rifles and automatic weapons, which blocked the only route over which the platoon could move. Knowing that it was the key to the enemy position, PFC Lobaugh volunteered to attempt to destroy this weapon, even though in order to reach it he would be forced to work his way about 30 yards over ground devoid of cover. When part way across this open space he threw a hand grenade, but exposed himself in the act and was wounded. Heedless of his

> wound, he boldly rushed the emplacement, firing as he advanced. The enemy concentrated their fire on him, and he was struck repeatedly, but he continued his attack and killed two more before he himself was slain.[120]

The Division then moved back to Aitape and then Hollandia. They would be part of the Reserve Force for the invasion of Leyte, MacArthur's quest to return finally fulfilled. Much more hard fighting awaited them.

120 https://www.32nd-division.org/old/history/moh/32moh.html, Accessed 14 December 2024.

CHAPTER VI

HOW TO DEFEAT JAPAN?

New Guinea was always the steppingstone for MacArthur's eventual return to the Philippines. We now for a brief period examine the strategic backdrop to what would become the SWPA's and the entire Pacific Theater's biggest campaign, the invasion of the Philippines (yes, it was bigger in numbers, resources, and time than Okinawa). A question debated at the time and one I fully embrace is the necessity of any campaign in the Philippines to win the war. The following pages are based on a previous book I authored, *Pacific Hurtgen. Pacific Hurtgen* detailed the plight of four U.S. infantry divisions, one of which was the 32nd, in northern Luzon during the spring of 1945. I questioned why those divisions fought battles within a campaign that was already won as I question why the Philippines themselves were actually attacked.

By the beginning of 1943, after the victories at Guadalcanal and Buna, the question on how best to defeat Japan began. The debate centered on three main possibilities:

1. The Navy, led by Chief of Naval Operations Admiral Ernest King and Commander of the Pacific Fleet Admiral Chester Nimitz, wanted a drive across the central Pacific, ending with an invasion of the island of Formosa. Occupation of Formosa would isolate Japanese forces in China and provide the perfect base from which to launch strategic air attacks against and eventually, invade Japan.
2. General MacArthur's preferred strategy of course meant a return to the Philippines. Citing America's honor, the fate of Filipino civilians, and the ability to interdict enemy supply lines, he also saw it as an ideal staging area for an eventual invasion of Japan.
3. Army Chief of Staff General George Marshall was in general

> agreement with Admiral King but would bypass Formosa as well as the Philippines and concentrate on Okinawa. Only 110 miles from Japan itself and not as exposed to enemy air power as would be Formosa or as distant as the Philippines, Marshall saw it as the perfect point to invade Japan.

This debate continued for nearly eighteen months. President Franklin Roosevelt made the ultimate decision. That decision was not to decide. In July 1944, a meeting with President Roosevelt, Nimitz, and MacArthur happened in Hawaii. Marshall, the President's most important advisor and someone he deferred to throughout the war, was not even there. FDR allowed both Nimitz and MacArthur to continue. Resources were dispersed rather than concentrated. Victory was inevitable but could have been accomplished far quicker and at less cost if only one was selected.

Which strategy would have worked best? The Philippines, Formosa, and Okinawa all offered excellent potential as invasion staging areas, with substantial ports and airfields. Land based tactical air power could not assist an invasion from either the Philippines or Formosa but could from Okinawa. Nimitz's aircraft carriers would have to pick up the slack. Marshall's approach required only limited ground forces to execute. Few of the Japanese garrisons of the various islands of the Pacific had large numbers of troops. Many were bypassed. Several Marine and Army divisions would be all that was needed to reach Okinawa, though an entire army was needed there. This took advantage of what was becoming the greatest naval force in world history.

The Pacific is vast. It dwarfs the Atlantic Ocean. To span the enormous distances in the Pacific required a massive fleet of not only combat vessels but support ships as well. It was only in the central Pacific that this force, produced at great expense and with unlimited capabilities, could be utilized. It is testament to General Marshall's greatness and selfless sacrifice that he saw this war in the Pacific would be won by the Navy.

General Marshall did not just dismiss an option for non-military reasons. He had misgivings about both the Philippines and Formosa. Formosa would require a massive invasion force and its proximity to China with millions of Japanese troops and hundreds of aircraft that could strike any invasion force was worrisome. Robert Ross Smith, the Army's official historian for the Philippine Campaign in 1945, commented on these concerns:

> Army planners quickly decided that Nimitz's new plans possessed major drawbacks. The Japanese would hardly allow Allied forces to sit unmolested in southern Formosa. Instead, the Japanese would mount strong counterattacks from northern Formosa with troops already on the island and with reinforcements staged in from China. Occupying and defending one beachhead on southern Formosa and another at Amoy would involve problems far different from those that the Allies had encountered previously, in the Pacific. So far during the war, the Japanese had usually been hard to move air and ground reinforcements against island perimeters that Allied amphibious task forces had seized. In the southern Formosa-Amoy area, on the other hand, the Allies would not have the protection of distance from major Japanese bases that they had enjoyed in earlier campaigns. The Allies did not have sufficient aircraft in the Pacific to keep neutralized all existing Japanese airfields within the range of southern Formosa and Amoy. In addition, the experience in the pacific had demonstrated that Allied air and naval forces could not be expected to forestall all Japanese efforts to move strong reinforcements across the narrow strait between Formosa and China.[121]

121 Robert Ross Smith, *Triumph in the Philippines,* 11-12.

Formosa would be forgotten or at least put on the backburner. With regards to the Philippines, Marshall continually reminded MacArthur, his subordinate (though one wonders if MacArthur ever considered himself anyone's subordinate), that strategy had to be dictated by military options. MacArthur had once held Marshall's job but did not seem to comprehend how Washington was making key decisions. He lectured MacArthur in a 1944 cable, reminding him: "We must be careful not to allow our personal feelings and Philippine political considerations to override our great objective, which is the early conclusion of the war with Japan.

The Navy's argument for a drive across the central Pacific was its immense fleet. The Pacific Fleet in the spring of 1944 had 535 ships, spearheaded by 15 large fleet aircraft carriers, dozens of light and escort carriers, hundreds of other combat vessels, and over 2,000 naval aircraft.[122] It all began on January 31, 1944, with the seizure of the Marshall Islands, setting up the seizure of the Mariana Islands. The Marianan islands of Saipan, Tinian, and Guam (a U.S. possession) were perfect sites for long-range bomber bases. The operation against the Marianas revealed to the Japanese high command how vulnerable the home islands would become, and they sortied their fleet to stop it. In history's second largest naval battle, the Battle of the Philippine Sea, Japanese naval air power was eliminated from the war. Several carriers and hundreds of planes were lost and those Japanese carriers that survived the battle returned home devoid of their aircraft. U.S. losses were negligible. Those planes lost were mostly because they ran out of fuel pursuing the retreating Japanese. MacArthur's New Guinea campaign was also wrapping up and the arguments over strategy started once again.

MacArthur's personality and ego figured prominently here. I believe, and this is solely a personal opinion, he considered his own personal glory as important as America winning the war. Perhaps he truly believed one could not be accomplished without the other.

---

122 Dan van der Vat, *The Pacific Campaign: The U.S. – Japanese Naval War 1941–1945*, (New York: Simon and Schuster, 1991), 317.

He could not allow the Navy, despite the merits of their arguments, to win the war. In January 1944 he pressed General Marshall to essentially make him the Eisenhower of the Pacific. When FDR appointed General Dwight Eisenhower to command Operation *OVERLORD* he was given command of ALL western European ground, air, and naval assets. The Army controlled both ground and air forces, the U.S. Air Force, as an independent branch of the military did not yet exist. An Army general had to be in charge. In the Pacific the Navy was clearly the dominant branch of the services. There is NO way Admirals King and Nimitz would give their ships to MacArthur. MacArthur told Marshall:

> I need only that which is now in the Pacific, it is sufficient if properly directed. I do not have to call on men or supplies being sent to the European front. That decision is one for the Joint Chiefs of Staff. I do not want command of the Navy, but must control their strategy, be able to call on what little of the Navy is needed for the trek to the Philippines. The Navy's turn will come after that. These frontal attacks by the Navy, as at Tarawa, are tragic and unnecessary massacres of American lives. The Navy operations proposed for the Marshall Islands would be the same. The Navy fails to understand the strategy of the Pacific, and fails to recognize that the first phase is an Army phase to establish land-based air protection so the Navy can move in.[123]

I'm sure the Navy would have also accused MacArthur of having no concept of successful strategy in the Pacific. Land-based air power was unnecessary when the Pacific Fleet could call on over 2,000 combat aircraft on the mobile launch platforms their carriers provided. Several vicious, protracted battles would be necessary to complete the drive across the central Pacific but these battles

---

123 Pogue, *Organizer of Victory, 1943–1945*, 440.

would be few in number. An invasion of the Philippines would require many protracted battles and cast the Pacific Fleet away to the role of spectators. Historian Stanley Falk, a critic of MacArthur's strategic designs and the motivations behind them, offers the following analysis:

> What MacArthur could not accept was that the Central Pacific offered the shortest and most direct route to bring early and decisive naval and air pressure to bear on the heart of Japan. Once they were available, the highly mobile carrier and amphibious forces, cruising freely across the open Pacific with their great floating logistical bases, would be able to advance further and faster and bypass enemy strong points far more readily than McArthur's and-based units.[124]

Militarily, MacArthur's arguments were valid. If a better option did not exist. Japan was dependent on Southeast Asia for the preponderance of their vital natural resource needs. Oil, rubber, metals—all traveled by ship northward along the Asian coast towards the Japanese home islands. Should that lifeline be severed Japan could be broken, possibly even without a direct invasion. The Philippines were the ideal location to interdict that supply line, a point MacArthur stated repeatedly.[125] The Philippines were also perfect to build bases or even enhance old American positions like Clark Field (vast air base on southern Luzon), bases from which an invasion of Japan could be launched.[126] As with Nimitz's invasion of the Marianas, the Japanese recognized how precarious their strategic situation became with American occupation of the Philippines. They sent the Imperial Navy, that which had survived the Battle of the Philippine

---

124 William M. Leary, *We Shall Return: MacArthur's Commanders and the Defeat of Japan*, (Kentucky: The University of Kentucky Press, 1988), 13.

125 Michael Schaller, *Douglas MacArthur: The Far Eastern General*, (London: Oxford University Press, 1989), 99.

126 Pogue, *Organizer of Victory, 1943–1945*, 443.

Sea, to contest the American landings on the Philippine island of Leyte in October 1944. The Battle of Leyte Gulf, history's largest naval battle and an overwhelming American victory, in some degree validates MacArthur's claim as to the Philippine's military value. Why else would the Japanese send all major units of their fleet, devoid of naval air cover, against the more powerful American Navy? MacArthur's forces were ashore, but again, was this step necessary? No, if there was also a drive across the central Pacific.

The argument that occupation of the Philippines would sever Japan's economic ties with Southeast Asia is superfluous, that connection was severed before the U.S. Sixth Army ever landed on Leyte. U.S. submarines savaged Japan's merchant fleet, depriving the military and home islands of many of their basic commodities, especially oil; 603 Japanese merchant ships were sunk in 1944. Luzon, the main Filipino island, was not even invaded until January 1945. These 603 ships totaling 2.7 million tons, over half of Japan's merchant fleet, also took 60% of Japan's bulk imports with them.[127] These submarines were based in Hawaii and remained so until the end of the war. Further, Okinawa was a far better base from which to invade Japan, only 350 miles from the Japanese home islands as compared to the roughly 1,200-mile trek from the Philippines. The Marianas served the same purpose as the Philippines and were already occupied with major air bases.

A Japanese army of 110,000 defended Okinawa. While undoubtedly a significant force it was barely one-third of the Japanese army in the Philippines or Formosa. MacArthur had conducted his own variation of an island-hopping campaign across New Guinea's northern coast during the spring and summer of 1944. He bypassed strongpoints and even cut off an entire Japanese army of 55,000 men when his forces landed at Hollandia in April. Rabaul, the powerful Japanese base on New Britain, north of New Guinea, was MacArthur's main objective until his New Guinea campaign began. It was a pow-

---

127 Van der Vat, *The Pacific Campaign: The U.S. – Japanese Naval War 1941–1945*, 375.

erful base with a significant defending force. Securing Rabaul would undoubtedly be a costly and lengthy process. Rather than subject his troops to a potential bloodbath, MacArthur's Fifth Air Force isolated and battered Rabaul to the point of uselessness. It was a brilliant campaign and one that demonstrated the flexibility and cunning of a great commander. Why was that not displayed here? The answer is simple: the recapture of the Philippines was personal to him and his superiors deserve equal blame for catering to him.

Whenever MacArthur's military arguments faltered he went to honor, both his own and that of the nation. The following dispatches tell that story:

> ... personal integrity, indeed my personal honor is involved if the Philippines were not a military objective.[128]

> ... the need to go back to the Philippines "where our unsupported troops" were destroyed by the enemy. We have a great national obligation to discharge. Moreover, if the United States should deliberately bypass the Philippines, leaving our prisoners, national and loyal Filipinos in enemy hands without an effort to retrieve them at the earliest possible moment we would incur the greatest psychological reaction.[129]

Subordinates also believed that his personal glory was a factor in his decisions. General William Gill, the 32nd's commander, commented:

> ... General MacArthur's great desire was to capture the Philippine capital so that his return would be dramatized for the natives and other people.[130]

---

128 Pogue, *Organizer of Victory, 1943–1945*, 444.

129 Ibid., 443.

130 Edward Jaquelin Smith, *Always A Commander: The Reminiscences of Major*

Admiral Nimitz also noted the MacArthur vitriol after a conference with him between March 25 and 27, 1944:

> Everything was lovely and harmonious until the last day of our conference when I called attention to the last part of the directive which required of him and me to prepare alternate plans for moving faster and along shorter routes towards the Luzon-Formosa-China triangle ... he (MacArthur) then blew up and made an oration of some length on the impossibility of bypassing the Philippines, his sacred obligations there—redemption of the 17 million people—blood on his soul deserted by the American people, etc., etc.[131]

When that wasn't enough, he inserted politics, which is in no way the purview of a military commander:

> I daresay that the American people would be so aroused that they would register most complete resentment against you (President Roosevelt) in the fall.[132]

Nor was he averse to ignoring the chain of command. When he felt General Marshall was not accepting his point of view he stated:

> ... that I be given early opportunity personally to present the case to the Secretary of War and to the President before finally determining my own personal action in the matter.[133]

He also believed that everyone was hampering his conduct of mil-

---

General William H. Gill, 82.

131 Charles A. Willoughby and John Chamberlain, *MacArthur: 1941–1951*, (New York: McGraw-Hill Book Company, Inc., 1954), 233.

132 Manchester, *American Caesar*, 369.

133 Pogue, *Organizer of Victory, 1943–1945*, 441.

itary operations. Once he had returned to the Philippines and the campaign was underway, he had most of his transports recalled. MacArthur was not happy about that:

> At this critical juncture (approaching Manila) I received a shocking order from Washington to release 70 of my transport ships at once. They were to return to San Francisco and be used to carry supplies and munitions to the Soviet forces at Vladivostok. I protested violently. The abrupt removal of these transports endangered the entire Philippine campaign and threatened the loss of thousands of our men fighting in north Luzon. No heed was given to my warnings. The government apparently had made up its mind that, in order to win the war against Japan, Russian help was needed. My views as to whether such a step at this late date in the war would be advisable were never solicited. As it turned out, all the supplies carried to Vladivostok by those ships, and hundreds of thousands of other tons, were eventually used in Korea and China against our own forces. So far as I know, not one shell, not one pound of food, not one stitch of clothing, and not one gallon of gasoline was ever used by the Soviet Army in the brief war they finally fought—if it can be called a war.[134]

The above-mentioned statement was written after the Korean War when MacArthur was undoubtedly trying to restore a reputation tattered by his removal from command. This was a paranoid man. I just don't see the rationale for such paranoia. If receiving an undeserved Medal of Honor was not enough, then no interference from above in any way affected his New Guinea campaign. He had his own air force, the Fifth, and his own naval support, the Sev-

134 MacArthur, *Reminiscences,* 244-5.

enth Fleet. Both were under his operational control. Despite a situation most of history's commanders would only hope to operate in, MacArthur saw enemies all around him. Michael Schaller, in *Douglas MacArthur: The Far Eastern General,* offers the following synopsis:

> MacArthur feared that institutional and personal enemies in Washington opposed his drive for ignorant or selfish reasons. To counter this opposition, he worked hard to sell his strategy and accomplishments to the American public and relied on the influenced of anti-New Deal political leaders and newspaper moguls. He alternatively cajoled and threatened, begged, and demanded so much that the President and Joint Chiefs committed greater resources to the Southwest Pacific than they thought wise. His wartime accomplishments were as much a testament to his public relations abilities as to his military skill.[135]

A definite ego was in place here. The personal aspect of this went beyond what he saw as elements undermining what MacArthur believed was the best way to win the war. There was a move towards public adulation, at the expense of subordinate commanders and the troops who fought and died in the Southwest Pacific. General Robert Eichelberger, along with Walter Krueger, was MacArthur's top ground commander. Eichelberger had saved the day at both Buna and Biak. Yet the public knew little of him. MacArthur ruthlessly controlled the press in his part of the war and Eichelberger, in a letter to his wife, explained why:

> ... (MacArthur) not only wants to be a great theater commander but he also wants to be known as a great frontline leader. This would be very difficult to put over if any of his particular leaders were pu-

135 Schaller, *Douglas MacArthur: The Far Eastern General,* 67.

> blicized. He leaves the impression with the people back home that he has been the one who has been doing the frontline fighting. This does not mean that he does not appreciate what I have done or that he does not give me a lot of mental credit. He just wants it all for himself. Unless one understands this dual feeling on his part of wanting to be a great strategic and also a frontline leader it will be impossible for anyone to understand the setup here.[136]

This appraisal was not limited to senior commanders. The following poem displays how many of his men viewed the SWPA commander:

> And while possibly a rumor now;
> Someday it will be fact
> That the Lord will hear a deep voice say
> "Move over, God – it's Mac."
> So bet your shoes that all the news
> The last great Judgement Day
> Will go to press in nothing less than
> Doug' Communique![137]

If there were no campaign in the Philippines, then America and the world would forget General Douglas MacArthur. He could not accept that and through his political connections and overt insubordination, a commander does not tell his superior that he wants to go over his head to the Secretary of War or even the President, he would remain relevant in the war. MacArthur was given far more latitude than any commander of the era to redeem his personal honor. Until his firing by President Harry Truman for insubordina-

---

136 Manchester, *American Caesar*, 360.

137 Schaller, *Douglas MacArthur: The Far Eastern General*, 73.

tion during the Korean War, he continued to push a personal agenda, often to the detriment of the nation's interests. Ironically, honor was the strongest part of MacArthur's argument, far stronger than his military points. But it was not his personal honor but the honor of the nation that offered the SWPA's commander's most compelling arguments. MacArthur stated in his post-war memoirs:

> I argued that it was not only a moral obligation to release this friendly possession from the enemy, now that it had become possible, but that to fail to do so would not be understandable to the Oriental mind. Not to do so, moreover, would result in death to thousands of prisoners, including American women, children, and men civilians, held in Philippine concentration camps. Practically all of the 17,000,000 Filipinos remained loyal to the United States and were undergoing the greatest privations and sufferings because we had not been able to support or succor them. To bypass isolated islands was one thing, but to leave in your rear such a large enemy concentration as the Philippines involved serious and unnecessary risks.[138]

The above-mentioned fact is a reality that many did not consider. MacArthur had spent many years in the Philippines prior to World War II and grew very fond of the Filipino people. Many of those suffering under Japan's iron fist were members of his command, the command he left in 1942. While he may have been egotistical and self-centered he did feel for those in the Japanese prisoner of war and concentration camps. Later in the campaign, when word came to him that an opportunity to save American prisoners presented itself, he jumped:

> I was deeply concerned about the thousands of prisoners who had been interned at the various

138 MacArthur, *Reminiscences*, 198.

> camps on Luzon since the early days of the war. Shortly after the Japanese had taken over the islands, they had gathered Americans, British, and other Allied nationals, including women and children in concentration centers without regard to whether they were actual combatants or simply civilians. I had been receiving reports from my various underground sources long before the actual landings on Luzon, but the latest information was most alarming. With every step that our soldiers took toward Santo Thomas University, Bilibid, Cabanatuan, and Los Banos, where these prisoners were held, the Japanese soldiers guarding them had become more and more sadistic. I knew that many of these half-starved and ill-treated people would die unless we rescued them promptly. The thought of their destruction with deliverance so near was deeply repellant to me.[139]

The famous Los Banos Raid followed immediately thereafter. The Filipino people and American prisoners were all in a desperate way. That was a valid argument for action. U.S. territory was attacked and occupied by the Japanese during World War II. Wake Island, Guam, Attu and Kiska in the Aleutians, and the Philippines were all victims of Japan's incredible success in the Pacific war's first six months. The basic duty of government, of all government, is to protect its people. Attu and Kiska had no military value and were only occupied by the Japanese to distract American attention away from their intended target of the island of Midway in June 1942. Yet, in the spring of 1943, both islands were invaded and secured. Attu was a very costly operation; Guam was retaken during the Marianas operation. Wake was handed back after Japan's surrender. The Philippines were American territory. Even if it made little sense militarily, and a far stronger argument existed for the

139 Ibid., 246.

Philippines than either the Aleutians or Wake, the American government has a legal and moral obligation to its people. This should have been all MacArthur stated. It may not have carried the day but was unassailable. Political threats and nonsensical paranoia are trumped by reality, both legal and moral. The thousands of American prisoners and millions of Filipino civilians living under the sword of Japanese brutality deserved a voice as well.

In the end, President Roosevelt did not decide. Both the drive across the central Pacific and MacArthur's return to the Philippines would continue. It is testament to the immense strength and resources of the United States that both campaigns were adequately supported and ultimately successful. A return to the Philippines was not as militarily decisive as the central Pacific drive. It should have been, as historian Dan Van der Vat stated: "... a one-horse race with a single jockey."[140] Nevertheless, MacArthur's personal honor, as he saw it, could now be redeemed. A significant Japanese army defended the Philippines. Its entire strength need not be confronted. Once the landings started on Leyte in October 1944, it was inevitable that Luzon, the main island of the Philippines and home to the capitol city of Manila would follow. A limited landing on Luzon would have achieved all the strategic purposes which MacArthur previously detailed when pushing for the operation. However, when MacArthur returned to the Philippines, he meant ALL the Philippines. No matter the strategic irrelevance, no matter the enemy strength, or the weakness of his own forces, the Japanese would not be permitted to control any Filipino territory.

---

140 Van der Vat, *The Pacific Campaign: The U.S. – Japanese Naval War 1941–1945*, 321.

CHAPTER VII

LEYTE—THE RETURN TO THE PHILIPPINES

FIGURE XIX. "He has returned!!!"

A famous picture, but completely fallacious. MacArthur came ashore after the beach was secure and fighting was already taking place inland. The nickname "Dugout Dug" in many ways came from this picture.

The landings in the Philippines were supposed to begin in January 1945. Macarthur's success in New Guinea and the lack of opposition as he continued moving north towards the Philippines convinced planners to move that date up. The major units of Nimitz's Pacific Fleet, with the Mariana operation complete in September, could shift support to MacArthur. On 20 October 1944, the U.S. Sixth Army landed on the Philippine island of Leyte.

Along with the landings Leyte is best known for history's largest sea battle. The Japanese sent the bulk of their still impressive surface fleet along with empty aircraft carriers (to draw away America's carriers and allow the Japanese to destroy the invasion forces) to Leyte and were decisively defeated. For the remainder of the war the Imperial Japanese Navy posed no significant threat to American forces.

The 32nd ID did not play any role in the initial landings. They remained at Hollandia as part of the Sixth Army reserve (along with the 77th Infantry Division). The landings were successful but costly. The Japanese were able to get reinforcements to Leyte from neighboring islands and fought as ferociously as ever. Their main point of entry was the port of Ormoc at the southern tip of the Ormoc Valley. They made their initial stand north of Limon. In what became the Battle of Breakneck Ridge, the 24th Infantry Division fought against several formidable Japanese units, included the recently arrived 1st Infantry Division. This division, considered one of Japan's finest, was the 32nd's main opposition on Leyte.[141]

For the 32nd Infantry Division the battle began with their landings near Pinamopoan, and the campaign progressed down Highway #2 through the Ormoc Valley towards Ormoc to the south. The Division would pivot west at the end of the campaign in the "mopping up" phase.

The 1st Division was a top opponent:

> This division was well equipped and adequately supplied. It had an organic field artillery regiment and in this campaign had available to it the supporting fires of another medium field artillery regiment. It had tanks and tankettes. Its infantry had been well trained and it lacked none of the tenacious, spiritual determination for which the Jap soldier is so highly lauded. This training was most evident by his selection, development, and utili-

---

141 32nd Infantry Division, *REPORT AFTER ACTION OF 32nd Infantry Division in Leyte Engagement of K-2 Operation for the period 16 November – 25 December 1944*, (Headquarters: 32nd Infantry Division, 1945), 5.

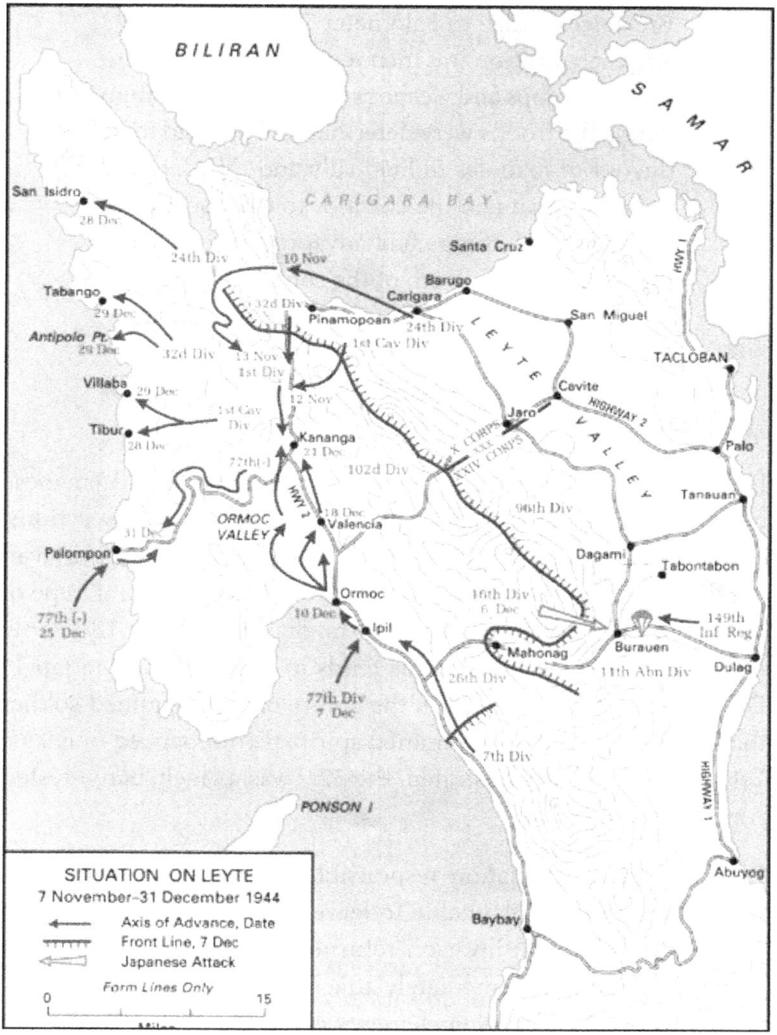

FIGURE XX. The Leyte Campaign

> zation of defensive terrain. These troops did not overlook a single strong defensive position from Breakneck Ridge to Kilometer 79 on Highway #2 and each position was manned with the proper number of troops and weapons to offer the maximum delay. The troops were determined. They had to be dugout of foxholes individually and this determination to maintain the corridor to the north coast even after our troops had advanced several thousand yards to the south of the southern end of the corridor. This training was also evident in his counter-intelligence measures. No Jap forces previously engaged by this division have been so thoroughly devoid of dog tags or personal identification.[142]

Further, this division was as tough and ferocious as all Japanese units. But it lacked the fanaticism the 32nd saw in previous actions, especially along the Driniumor River. As stated by one American officer: "There was less evidence of the spiritual or 'Banzai' type of warfare which was so prevalent in the divisions of the 18th Army recently engaged at Aitape. This tends to indicate that the Jap 1st Division carried more faith in the ability of a well-trained soldier than in the ability of an untrained spiritually influenced one ..."[143] Yet, as the battle demonstrated, the 32nd was a crack, battle-tested American unit.

X Corps, the organization responsible for the drive to Ormoc, needed help. The 32nd, unable to leave Hollandia until the transports from the initial invasion returned, landed on Leyte on 14 November and was immediately attached to X Corps. By 16 November they were relieving elements of the 24th ID on Breakneck Ridge. For the first time the entire Division would fight as one. The

142 Lt. Colonel Leonard S. Carroll, *Destruction of Japanese 1st Division in the Leyte Campaign*, (Headquarters: United States Army Forces In The Far East, 22 February 1945), 3.

143 Ibid.

Division's After-Action Report sums up their time on Leyte:

> ... During the next five and one-half weeks, or until 25 Dec., the 32d Div. fought an aggressive offensive action against the well-trained and amply equipped Jap 1st Div. This Jap Div., some 12,000 strong when it landed, was completely defeated by the end of a five-week period. 6400 members of it were counted killed, 37 PDW taken, a total of 20 light tanks, 7 tankettes, 45 artillery pieces, 205 machine guns and huge stocks of ammunition captured or destroyed. The Jap division's final stand, made with some 500 defeated, exhausted troops (all that the Divisional Commander had been able to reassemble); was reduced and the troops driven from strong defensive positions one mile S of BONBONGON on the 22nd of Dec. The action was characterized by close hand-to-hand combat in densely jungled, mountainous terrain, during torrential rainfall throughout the period.[144]

The 128th IR was the first Division regiment committed to action.

The 3rd Battalion of the 128th encountered stubborn resistance from the Japanese beyond Breakneck Ridge but by 21 November overcame it and the unit reached the Leyte River. Spider holes, caves, and jungle-covered bunkers were blown up with satchel charges, grenades, and bazooka teams. The days of Buna and inadequate weapons now a distant memory. Along with their sister battalions, movement was slow but steady and the regiment reached the high ground east of Limon, forcing the Japanese in the area into an ever-constricting pocket. The Japanese held out until 12 December. It took over three weeks for an advance of about 2,000 yards. How-

---

144 32nd Infantry Division, *REPORT AFTER ACTION OF 32nd Infantry Division in Leyte Engagement of K-2 Operation for the period 16 November – 25 December 1944*, 6.

ever, 850 soldiers were killed and the Japanese 1st Division's 57th Infantry Regiment destroyed.[145]

PTO terrain was often as formidable an enemy as the Japanese. Maintaining contact between units grew ever harder. Gaps appeared that required increased patrol activity to secure. Such a gap existed between Limon and HILL 1525. The 126th IR was dispatched to HILL 1525, occupying it on 23 November, encountering only scattered resistance enroute.

For the 126th, their entry into battle on Leyte did present an unanticipated problem. They had no knowledge of the terrain and enemy they faced. Their assignment, as stated by the unit AAR, "... was the spur of the moment decision of higher headquarters assigning an entirely different zone than had previously been contemplated."[146] They were unable to conduct any proper reconnaissance, and no aerial photos existed. Further, poor maps, an affliction that plagued not only the 32nd but also all formations employed in the Philippines, made the 126th's job difficult.

The 126th had at times a busy and at other times quiet campaign. It started in good shape physically, 2,276 officers and men available for action.[147] Arriving on the 19th of November they launched several attacks on 20 and 22 November in the area of Hill 1525. The remainder of November proceeded as follows:

- 23 November – Quiet
- 24 November – Patrols, more problems with inaccurate maps
- 25 November – Cleared a trail of numerous Japanese machine gun nests
- 26 November – Little activity
- 27 November – Seized a Japanese occupied ridge

145 Ibid., 7.

146 126th Infantry Regiment, *SUPPLEMENTARY REPORT AFTER ACTION, 26 Dec. 44 – 24 Jan. 45*, (Headquarters: 126th Infantry Regiment, 1945), 3.

147 Ibid., 1.

28 November – Little activity, patrolling

29 November – Little activity, patrolling

30 November – Little activity, patrolling

Despite the frequent days of quiet during this week losses were not light: 50 KIA, 148 WIA, 7 MIA.[148]

From Hill 1525 the Regiment was ordered to occupy the high ground between the Leyte River and Highway #2, north of Bonbongon. Enroute several Japanese strongpoints were eliminated. Once on the high ground roadblocks were put up along Highway #2 to prevent Japanese reinforcements from reaching the area. The 126th conducted extensive patrolling but faced a new problem: dwindling supplies. Supplies coming forward by truck were limited by the poor terrain and occasional ambushes and enemy sniper fire. Airdrops picked up the slack, though their inaccuracy still left shortages. Food had not been delivered since 26 November and by 2 December the Regimental history noted: "... the men were becoming weak with hunger."[149] Enemy forces, including Japanese tanks, repeatedly attacked these roadblocks. Artillery and mortar fire and the lethal bazooka teams broke up every attack. Private First Class John Dirk Vlug of the 126th Infantry Regiment's 1st Battalion, was one of those lethal bazooka men. He received the Medal of Honor for his actions on 15 December. He acted alone, not as a member of a bazooka team, making his actions even the more remarkable. The Citation reads:

> ***... When an American roadblock on the Ormoc Road was attacked by a group of enemy tanks. PFC Vlug left his covered position, and with a rocket launcher and six rounds of ammunition, advanced alone under intense machinegun and 37-mm fire. Loading single-handedly, he destroyed the first tank, killing***

148 126th Infantry Regiment, *REPORT AFTER ACTION: LEYTE OPERATION,* (Headquarters: 126th Infantry Regiment, 1945), 6-8.

149 Ibid., 9.

its occupants with a single round. As the crew of the second tank started to dismount and attack him, he killed one of the foe with his pistol, forcing the survivors to return to their vehicle, which he then destroyed with a second round. Three more hostile tanks moved up the road, so he flanked, the first and eliminated it, and then, despite a hail of enemy fire, pressed forward again to destroy another. With his last round of ammunition, he struck the remaining vehicle, causing it to crash down a steep embankment. Through his sustained heroism in the face of superior forces, PFC Vlug alone destroyed five enemy tanks and greatly facilitated successful accomplishment of his battalion's mission.[150]

On 4 December, a patrol with several bazooka teams knocked out nine Japanese tanks in a single day.[151] Every night saw the enemy trying to infiltrate American positions or launching desperate banzai attacks.

Japanese banzai attacks, suicidal charges that stood little chance of success, were a common event in the PTO. Masses of Japanese soldiers would leave their positions screaming in a human wave attack. The officers waved their swords and the soldiers that followed, often out of ammunition, followed their leaders. These fanatical attacks stood little chance of success because American units, especially a veteran unit such as the 32nd ID, had established doctrine for this. The 127th IR noted: "If the terrain permits, hold for the night no later than 1600 so that a complete reconnaissance of the area may be made to set up a strong perimeter in order to be prepared for the inevitable counterattack."[152]

150 https://www.32nd-division.org/old/history/moh/32moh.html, Accessed 17 December 2024.

151 126th Infantry Regiment, *REPORT AFTER ACTION: LEYTE OPERATION*, 10.

152 127th Infantry Regiment, *History – Leyte – 127th Infantry Regiment – 32nd In-*

FIGURE XXI. Infantry in the jungle of Leyte

This was typical terrain on Leyte. Jungle, perfect camouflage, and inaccessible to all but foot soldiers.

FIGURE XXII. Knocked out Japanese tank

A victim of the outstanding 32nd Infantry Division's bazooka teams. Japanese tanks were small, lightly armored, and with limited firepower. Bazookas could easily destroy them.

The ground allowed these continual attacks by Japanese tanks and infantry as the highway was a continuous series of turns that made observation difficult and the thick foliage off the road allowed Japanese infantry to often infiltrate within yards of the roadblocks. The road, as described by the 32nd AAR, displays the problems faced:

> This road is extremely crooked as it winds around the slopes of each individual nose of the hills. These hills are densely wooded and the Japanese were able to move off the road and over a hill, which would be beyond our vision because of the crookedness of the road. Moving through the forest, he would again come onto the road, cross it into more densely forested terrain and move on to the N or S. Our patrols contacted and killed numerous Jap parties attempting to move through this area.[153]

Kilay Ridge was the Division's next objective.

Kilay Ridge was part of a series of ridges that dominated the terrain west of Highway #2. As an artillery observation point it was vital and the Japanese repeatedly tried to drive the 1st Bn., 34th IR, 24th ID off.[154] The 2nd Bn. of the 128th IR relieved the 1st Bn. of the 34th on 2 December and immediately attacked the enemy forces that had harassed the American defenders on Kilay Ridge. Artillery devastated the Japanese 1st Division's 2nd Bn., the Ridge proving its value as an observation point. 200 Japanese soldiers were killed and large quantities of weapons and equipment captured.[155]

---

fantry Division, 6 November 1944 – 5 January 1945, (Headquarters: 127th Infantry Regiment, 1945). "Lessons Learned."

153 32nd Infantry Division, *REPORT AFTER ACTION OF 32nd Infantry Division in Leyte Engagement of K-2 Operation for the period 16 November – 25 December 1944*, 9.

154 Ibid., 10.

155 Ibid.

FIGURE XXIII. Highway #2

Though called a highway, this was little more than a muddy road with countless twists and turns. U.S. heavy equipment such as tanks found most of the "Highway" unusable.

Though the Japanese 1st Division was considered one of its best it did not mean it still could not devolve into what became the more frequently encountered enemy ground force. Throwing away men for no tactical purpose happened on Kilay Ridge:

> The most noteworthy example of his dissipation of troops in piecemeal efforts was in the Kilay Ridge area. We occupied Kilay Ridge with a battalion. Kilay Ridge was a bald, open hill and he knew what strength we had on it. Yet with the entire 1st Infantry Regiment at his disposal he threw away the entire 2d Battalion of that regiment in repeated attempts to climb the ridge through our heavy artillery concentrations. The other glaring example was the push to the north coast by the 2d Battalion of the 49th Infantry. He pushed the battalion through but when it got into the hot water he let stew, rather than assisting its withdrawal with the other two battalions of the 49th, which were in that area.[156]

---

156 Lt. Colonel Leonard S. Carroll, *Destruction of Japanese 1st Division in the Leyte Campaign*, 4.

The Japanese also failed to use their artillery effectively, except in isolated instances, throughout most of the campaign. Harassing fires replaced observed, coordinated barrages and "huge" amounts of ammunition remained unused, lying along Highway #2 after the Japanese withdrew.[157]

The final phase of the operation, the destruction of the remaining Japanese units on Leyte, would begin on the night of 5 December. The 32nd's initial role in this operation was for the 126th and 127th IRs to seize the high ground running through Lonoy. Approaching the Leyte River Bridge they encountered what the Divisional history called: "... one of the strongest defensive positions of the campaign. Numerous foxholes and spiderholes 10 feet deep, in cases, were connected by interlacing communications trenches and all of the positions were extremely well camouflaged."[158] Two days of heavy, close-in fighting followed where 183 Japanese soldiers were killed and ten pieces of heavy artillery captured.[159] For the next four days, 8-11 December, the 126th and 127th IR's fought enemy forces occupying the high ground west of Highway #2.

5 December saw another of the Division's three Medals of Honor conferred to Private First Class William A. McWhorter of M Company, 126th IR. PFC McWhorter:

> *... PFC McWhorter, a machine gunner, was emplaced in a defensive position with one assistant when the enemy launched a heavy attack. Manning the gun and opening fire, he killed several members of an advancing demolition squad, when one of the enemy succeeded in throwing a fused demolition charge in the entrenchment. Without hesitation and with complete*

157 Ibid.

158 32nd Infantry Division, *REPORT AFTER ACTION OF 32nd Infantry Division in Leyte Engagement of K-2 Operation for the period 16 November – 25 December 1944*, 11.

159 Ibid.

> *disregard for his own safety, PFC McWhorter picked up the improvised grenade and deliberately held it close to his body, bending over and turning away from his companions. The charge exploded, killing him instantly, but leaving his assistant unharmed.*[160]

This was not the last acknowledged act of incredible heroism by soldiers of the Division during this campaign.

The fights against the enemy forces encountered on Leyte were no different than those previous when the Japanese occupied defensive positions. Unlike Buna, on Leyte the Japanese had ample heavy weapons and ammunition and, in some locations, had several years to prepare and improve their positions. Attacks by the 32nd, the most veteran of all the Army units in the PTO, tended to follow the pattern below:

1. Patrols locate an enemy position.
2. Artillery and mortar fire blankets the enemy position.
3. If observable from the air, which was rare in the jungle terrain of Leyte, air attacks preceded the ground attack.
4. Infantry, usually no more than 50 yards behind mortar and artillery fire, moves into position to assault the enemy position.
5. Trained assault teams destroy the position.

The above worked. However, it was time consuming and often costly. The assault teams were usually part of an organic infantry platoon. An infantry platoon can have 35-38 men at full strength though as battles progressed it was rarely the case. An infantry platoon in the 32nd ID SHOULD look like what follows:

Platoon Leader (2LT)

Platoon Sergeant (SFC)

160 https://www.32nd-division.org/old/history/moh/32moh.html, Accessed 29 December 2024.

Rifle Squad Leader (SSG) - 9 riflemen, 2 Browning Automatic Rifles (BAR's)

Rifle Squad Leader (SSG) - 9 riflemen, 2 BAR's

Assault Team Leader (SSG) - 1 or 2 flamethrower teams (2-4 men), 1 bazooka team (2 men), 1 30-caliber machine gun team (2 men), 2 explosives men (satchel charges, bangalore torpedoes, etc.)

The numbers and composition of the assault team varied depending on the situation and availability of men and weapons. The Company's Weapons Platoon could provide additional machine gun, flamethrower, and explosive teams. Nor should the ranks of the various leaders within the platoons be considered ironclad. With each attack casualties mounted. Sergeants became Platoon leaders, Corporals Squad leaders. The rifle squads provided covering fire as the assault teams moved in. To use a flamethrower a team would have to get within 50 feet, a very vulnerable position. The explosives men had to get even closer. Usually successful, they moved on to the next position. By 11 December they had destroyed dozens of positions, but further advances were painstakingly slow. Assault teams and the tactics previously described were not always possible. Highway #2 continued to frustrate the advance for a variety of reasons:

> ... the 126th had advanced a short distance to the S and was meeting opposition on both sides of the road. Every bend of the road was lined with Jap foxholes dug into the banks of the road and spider holes dug underneath the roots of trees and under logs on the hillsides. It was bitter, close, hand to hand fighting and because of the steepness of the terrain, the denseness of the tree growth, the inaccuracy of maps and nearness of adjoining units, artillery and mortar fire could not be used to its full advantage in reducing these positions. Medium tanks were used, however, and accompanied

> the leading elements on Hwy 2 at all times. Direct fire from the 75mm guns on the tanks was effectively utilized in reducing enemy positions on or adjacent to the road but fire on targets some distance off the road was limited by same restrictions as artillery.[161]

This would be the type of fight for the rest of the campaign.

The Division continued moving south down Highway #2 and again met heavy opposition. Two ridges overlooking the Highway from the east were covered by plentiful automatic weapons. Often in the war Japanese positions were last stands, the men occupying them expecting to fight and die there. Little consideration was attached to mutually supporting positions. That was not the case here (and would often not be the case on Luzon) as the two ridges supported each other. 32nd troops could not flank or maneuver to the rear without facing an intense crossfire.

Sgt. Leroy Johnson led a patrol on 15 December to try and find a way to secure one of those ridges. His remarkable actions earned him the third of the Division's Medals of Honor for the Leyte Campaign:

> ***... Sgt. Johnson was a squad leader of a nine-man patrol sent to reconnoiter a ridge held by a well-entrenched enemy force. Seeing an enemy machine-gun position, he ordered his men to remain behind while he crawled to within six yards of the gun. One of the enemy crew jumped up and prepared to man the weapon. Quickly withdrawing, Sgt. Johnson rejoined his patrol and reported the situation to his commanding officer. Ordered to destroy the gun, which covered the approaches to several other enemy***

161 32nd Infantry Division, *REPORT AFTER ACTION OF 32nd Infantry Division in Leyte Engagement of K-2 Operation for the period 16 November – 25 December 1944*, 11.

> ***positions, he chose three other men, armed with hand grenades, and led them to a point near the objective. After taking partial cover behind a log, the men had knocked out the gun and begun an assault when hostile troops on the flank hurled several grenades. As he started for cover, Sgt. Johnson saw two unexploded grenades, which had fallen near his men. Knowing that his comrades would be wounded or killed by the explosion, he deliberately threw himself on the grenades and received their full charge in his body. Fatally wounded by the blast, he died soon afterward.***[162]

A further advance down Highway #2, to bypass the ridges, was also not possible as the ridges commanded the route. Tanks joined the fight but a series of Japanese roadblocks (disabled Japanese light tanks) limited the advance. Infantry from the 127th IR arrived but could do little because of the heavy fire from the ridges. The infantry pulled back and during the night of 21 December every available gun and mortar pounded the ridges. The heavy fire worked as the following morning the advance happened without incident. The Japanese had abandoned the ridges. This, as stated by the Divisional history, "... marked the collapse of the Japanese resistance in the upper ORMOC VALLEY."[163]

The tanks used during these attacks on the ridges could have operated more effectively had they trained with the infantry they fought with more extensively beforehand. On 19 December a platoon of Sherman tanks was attached to the 127th IR, a unit it had not trained with. Several tanks and many infantry were lost to enemy fire because neither the tankers nor the infantrymen knew how to protect the other. For a planned 21 December attack a new tank

---

162 https://www.32nd-division.org/old/history/moh/32moh.html, Accessed 2 January 2025.

163 32nd Infantry Division, *REPORT AFTER ACTION OF 32nd Infantry Division in Leyte Engagement of K-2 Operation for the period 16 November – 25 December 1944*, 12.

platoon from the 763rd Tank Battalion arrived. They had worked with the 127th before. With infantry on either side two tanks (all that could move down the road at any one time) moved down the road. The infantry selected the targets and kept the Japanese from employing their usual bag of tricks against the tanks (mines, satchel charges, sticky bombs).[164] General MacArthur became famous, according to his supporters, and infamous, according to those who did the actual fighting, for his public proclamations that a campaign was over and only "mopping up" remained to be done. He had done so at Buna and again at Biak several months earlier. For those doing the "mopping up" the harshest fighting often awaited them. It meant digging the Japanese, who at that point of a battle knew they had lost but would not surrender and were more than happy to die for their Emperor and take as many Americans as possible with them, out of their spiderholes, bunkers, and caves. On Leyte that job fell to the 128th IR. Note again MacArthur declared the campaign concluded on 24 December. The Regiment's AAR described what they faced while "mopping up:"

> On 24 December, General MacArthur announced the Leyte Campaign had been concluded except for mopping up. However, for us, this mopping up meant what turned out to be an eight-day march over some of the most difficult mountain trails we had encountered since our training days in Australia prior to the Buna campaign. The 1st and 2nd Bns were selected to make the march. The 1st Bn and two platoons of Anti-Tank Co on the left and the 2nd Bn with Regimental Hq and Hq Company, detachment of Service Company for supply purposes, Cannon Company and detachment of Anti-Tank Company on the right. The Bns attempted to follow former trails about 600 to 1000 yards apart.

---

164 Committee 16, *Armor On Leyte*, (Fort Knox, Kentucky: The Advanced School, Officers Advanced Course, 1949), 57-8.

> This was done most of the trip. On 24 December the going was very hard with steep ridges all along, slippery muddy trails hindered our advancement ... During the day the Regiment killed 27 Japs. On 25 December, the 2nd Bn group saw the hardest marching. The morning was spent in climbing to the top of a mountain ridge. The climbing was difficult but as we later found out, the descent was much worse. The trail led almost perpendicularly down the side. After reaching the bottom, another ridge was encountered, this, almost straight up, everyone had to use handholds to pull themselves up. All in all there were seven ridges from the bottom of the first descent to the first bivouac area ... The 1st Bn had encountered the same difficult going. In the early morning the 1st Bn closed in on an estimated 300 to 400 Japs armed with HMG's and LMG's ... They routed the Japs driving them to the West. Artillery fire was directed on the fleeing Japs, but due to the heavily wooded terrain observation was difficult. The 1st Bn killed 53 Japs and we lost one officer killed. On 26 December, both Bns rested and were resupplied by artillery cub plane ... Om 28 December, the Regiment could not move as we were out of rations and had to wait for a ration drop by C-47's ... On 30 December the Regiment reached its destination.[165]

Yet what accolades did this Regiment receive? It's as if they were forgotten because they "mopped up." I only mention accolades because the 127th IR received War Department citations, deservedly so, for their outstanding actions on Leyte. The 3rd Bn of the 127th was cited in an official War Department order for actions per-

165 128th Infantry Regiment, *HISTORY: LEYTE,* (Headquarters: 128th Infantry Regiment, 1945), 13.

formed from 30 November through 7 December. Company C of the 127th also received a Distinguished Unit Citation (See Appendix III) for their actions on Highway #2 on 15 December. The 126th IR received individual accolades for acts of heroism. Three Medals of Honor were awarded to the Division for Leyte. The 126th received all of them, as well as two of the three Distinguished Service Crosses (the other awarded to a member of the 127th). They all received the congratulations and admiration of General Gill, their Division Commander. On 22 December, before MacArthur declared the Leyte operation over, he issued the following General Order #104 to all the men of the Red Arrow Division:

> ***Today the "Red Arrow" Division successfully completed its primary mission of forcing a passage through the mountains from Pinamopoan to the Ormoc Valley. After thirty-six days of the bitterest hand-to-hand fighting yet experienced in this war the Division has annihilated the 1st Imperial Division (reinforced), and by this determined action has shortened the completion of the Leyte Campaign.***
>
> ***Every officer and every enlisted man in the Division as well as those attached played a vitally important part in the Division's success.***
>
> ***I wish to compliment each individual and to express my personal appreciation for the splendid work accomplished by them in this campaign. Without this coordinated effort of each individual the Division could not have been successful.***
>
> ***I extend the Season's Greetings to each of you, and in so doing, express my confidence in your continued success. May God watch over you and help you through the strenuous days ahead.***[166]

166 https://www.32nd-division.org/, Accessed 4 January 2025.

The cost to the Division for Leyte was substantial: 450 KIA, 1491 WIA, 8 MIA.[170] The infantry regiments suffered most of the casualties. They had just settled in for a rest when they received orders to be ready to move again in two weeks. No replacements were received nor were equipment losses made good. This had been a quick fight but it did take its toll. Far worse lay ahead.

CHAPTER VIII

THE VILLA VERDE TRAIL, PART I

This was a mauling fight against the Jap in his remarkable defensive positions, against the terrain, supply, and climate. In those 119 days, the Red Arrow boys fought 22 miles, sometimes 35 yards at a time, with the Jap never more than 30 feet away. The Division killed 9,000 Japs and took 50 prisoners. It lost 4,226 men, about a third of the Division strength.

—John M. Carlisle, *Red Arrow Men: Stories About the 32nd Division on the Villa Verde*

After finishing their work on Leyte the 32nd was destined for Luzon. Luzon, the largest Philippine island and home to the capitol city of Manila, had the largest army U.S. forces encountered during the war. Approximately 260,000men divided into three groups under the overall command of General Tomoyuki Yamashita, defended Luzon. The Kembu Group, 30,000 strong, defended the vital position of Clark Field, the island's largest air base and the Bataan Peninsula. The Shimbu Group's 80,000 men deployed primarily around Manila. The largest contingent, the Shobu Group, 152,000 strong and under Yamashita's direct command, anchored itself in the Cagayan River Valley and the mountains of northern Luzon. It was a formidable force that had stockpiled weapons and ammunition while preparing defensive positions for three years. The Sixth Army's I Corps (25th, 32nd, and 33rd Infantry Divisions) received the mission to destroy the Shobu Group, a mission far beyond their capabilities.

FIGURE XXIV. The Philippine Islands

Leyte, attacked first, set the stage for Luzon, the main objective. Luzon had the largest Japanese army in the Pacific, outside of China.

General Yamashita was a smart, seasoned commander. Despite the size of the Shobu Group he knew his army's limitations. They lacked transport, air support, communications equipment, and other vital supplies. They were not an offensive force in any way. Yamashita knew he was a sacrificial force.[167] He had to protect the Cagayan Valley, the main source of food for his army, and turned all he could into fortresses. Some were mutually supporting, especially on the Villa Verde Trail, others were not. They only became relevant if attacked. Standard U.S. Army doctrine called for a 3 to 1 numerical advantage when launching an attack. The Japanese outnumbered I Corps by over 2 to 1, though that Japanese strength was scattered. It begs the question, why attack? At this point, General MacArthur reenters the discussion.

The landings on Luzon began on 9 January 1945 in Lingayen Gulf. By the end of February Manila (destroyed in the process with over 100,000 dead civilians) and Clark Field were secure. This gave MacArthur the port, airfields, and staging areas he needed to prepare for the invasion of Japan, an invasion he believed he would lead. The 38th and 43rd Infantry Divisions attacked the Bataan Peninsula. This attack protected Manila from the south. The I Corps, staged north of Manila, became, or should have become, a covering force. Given Yamashita's stationary Group the three divisions, including the 32nd, were more than enough, with plentiful artillery support, to keep the enemy far north. MacArthur did not accept this situation. He wanted every inch of the Philippines liberated and ALL enemy forces destroyed. Several divisions were taken from Krueger's Sixth Army to form a new organization, the Eighth Army, under the command of General Eichelberger. The Eighth Army, throughout the coming spring, launched operations against the tactically and strategically valueless islands and positions of the southern Philippines. The I Corps would fight and toil, without real rest or reinforcements, against the Shobu Group.

---

167 Robert Ross Smith, *Triumph in the Philippines*, (Washington D.C.: Center Of Military History, 1963), 90.

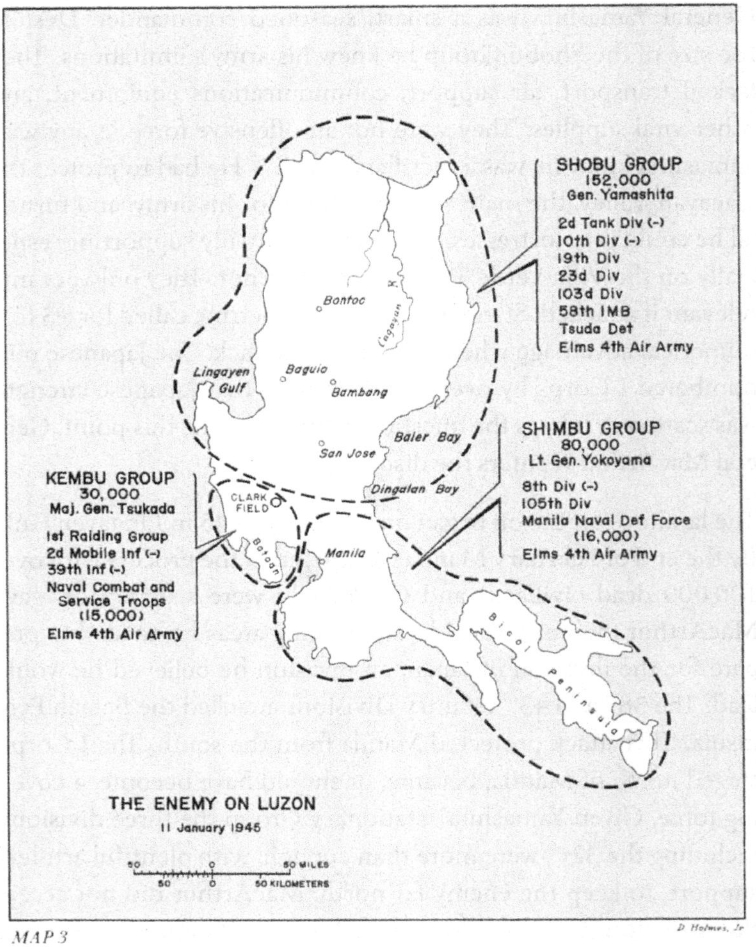

FIGURE XXV. Japanese Strength on Luzon

Northern Luzon, where the 32nd Infantry Division would spend the next 6 months, had the largest portion of Yamashita's strength.

The I Corps three divisions would each move on a different route towards Santa Fe, a gateway to the Cagayan Valley. The 33rd ID drove down Route 11 towards the town of Baguio. The 32nd, the center division of the I Corps, would move down the Villa Verde Trail towards Santa Fe, hoping to join the 25th ID, fighting its way to Santa Fe through the Balete Pass. Of the three divisions the 33rd met the least resistance. The 25th finished the operation battered and exhausted. The infantry division that saw more combat than any other American unit in World War II, was physically and emotionally finished by the end of June.

The rationale for this operation escapes me. The Sixth Army After Action Report for the Luzon operation stated: "... the Japanese commander expected that his forces defending the important strategic areas on Luzon would make U.S. troops pay a very high price for every advance and intended to launch the forces he had assembled on the northern edge of the Central Plain at the proper time in a coordinated counter-offensive against the Sixth Army's left."[168] This is ridiculous. The Japanese did not have the capability, transport, air power, or anything else to launch such a counter-offensive. Yamashita's Shobu Group was scattered over hundreds of square miles of land, lacking the communications equipment to launch such an attack. Yamashita's goal was to delay the inevitable and make the American forces come to him. Launching a major attack negated any advantage he might have had. This basis to send I Corps north rather than just leaving the Japanese there to wither away is shameful.

When the 32nd began the Leyte operation the Division was at full strength. Losses from the Driniumor River campaign were replaced. Such was not the case when they landed on Luzon. With little more than 11,000 officers and men they were 4,000 below their authorized strength. One-third of that number had been with the Division at Buna.[169] They were tired and the rest, as little as two weeks, they received after Leyte did not help. Much of that time went to reorganization and preparing for the landings on Luzon. Though this time of rest was brief it was the only time replace-

---

168 Sixth Army, *SIXTH UNITED STATES ARMY: REPORT OF THE LUZOIN CAMPAIGN: 9 JANUARY 1944 - 30 JUNE 1945, VOLUME I*, (Headquarters: Sixth Army, 1945), 79.

169 Smith, *Triumph in the Philippines*, 504.

ments could be absorbed and integrated into front-line units. Unfortunately, that did not happen. The first wave of replacements did not reach the 32nd until March and were thrust into battle green.

The 11,000 number of men available for the 32nd at the beginning of this campaign is actually worse than it appears. One must understand the composition of the World War II American infantry division. The authorized strength of 15,000 seems impressive. Not all of those 15,000 were actually infantry. The foundation of the infantry division was its three infantry rifle regiments, each with an authorized strength of 3,118 men for a total of 9,354.[170] Again, it sounds great but even that number is misleading. What did the other 6,000 men of the infantry division do?

Below are the other formations within the prototypical American infantry division:

Artillery Regiment

Headquarters Company – various staff section
(division nerve center)

Combat Engineer Battalion

Medical Battalion

Cavalry Reconnaissance Troop

Ordnance Company

Quartermaster Company

Signal Company

Military Police Platoon[171]

None of the other organic division organizations contained any riflemen. That did not mean they were not necessary. They were vital. Engineers carved roads out of dense jungle and mountains. They

170 Robert Young, *Pacific Hurtgen,* (Washington D.C.: Westphalia Press, 2017), 48.

171 Ibid.

dug bunkers. The Medical Battalion, Ordnance, Signal, and Communications companies all served an important purpose. Especially in the environments of the PTO, artillery was the infantry's best and most reliable friend. The Artillery Regiment had 54 105-mm. and 12 155-mm. Howitzers. Additional 105 and 155-mm. Guns as well as huge 8- and 9-inch howitzers could augment those numbers. Even in the PTO the infantry division had their normal haul of vehicles: 13 M8 armored cars, 5 halftracks, and 1,371 other vehicles (jeeps, trucks, etc.).[172] Due to terrain restrictions and often non-existent road nets, those 1,371 vehicles often stayed behind. Independent tank battalions provided armored support. Engagements with Japanese tanks were a rarity. In the PTO the tank was an infantry support weapon.

The standard infantry regiment had three rifle battalions. The regiment's 3,118 personnel broke down as follows:

> Infantry Battalions (3) – 860 men per battalion
>
> Headquarters Company – 108 men (service personnel, Communications Platoon, Reconnaissance Platoon)
>
> Anti-tank Company – 165 men, 9 towed 57-mm. Anti-tank guns, Minesweeping Platoon
>
> Cannon Company – 118 men, 6 105-mm. Self-propelled howitzers
>
> Service Company – 115 men, staff sections and Transportation Platoon
>
> Medical Detachment – 135 men to include doctors, aids and medics[173]

I have poured through the records of the 32nd Infantry Division and many other infantry divisions for different projects. I have rarely found any mention of the Anti-tank companies. Towed anti-tank

---

172 Ibid.

173 Ibid., 49-51.

guns were a short-range weapon that would have to be dragged through mud and jungle to an effective firing position. Artillery was a long-range weapon and thus far more effective. Could Anti-tank companies not be replaced by an additional rifle company?

With 860 men per rifle battalion that would seemingly give a strength of 2,580 infantry. Again, the number is misleading. Only 916 of that 2,580 were riflemen. Those infantry battalions had within their 860 men 3 rifle companies, a heavy weapons company, and a headquarters company. The Headquarters Company had the usual service, support, and transport platoons. The Heavy Weapons Company had 6 81-mm. mortars and 8 30-caliber machine guns. Each rifle company, with a strength of 180 men, had 3 rifle platoons, a headquarters element, and a weapons platoon. The rifle companies had 305 total infantrymen.[174] Why all this? The amount of foot soldiers available for the coming fight was far too low.

Each of the Division's three infantry regiments landed on Luzon with a strength of approximately 2,000 men. Over 1,000 understrength.[175] The 128th Infantry Regiment was the most depleted with only 1,500 effectives, less than half their authorized strength. The infantry battalions were of course the hardest hit on Leyte and as previously mentioned, those losses had yet to be made good. The Division would be in continuous combat for the next five months. Never were they at full strength or peak efficiency.

The 32nd had dealt with the swamps of Buna, the jungles of the Driniumor River, and the mountains of Leyte. Luzon and its Villa Verde Trail saw the hardest fighting and worst conditions yet experienced:

> The terrain in this area was much worse than any that the Division had so far encountered. Hills with nearly perpendicular slopes and deep, pre-

174 Ibid.

175 Sixth Army, *SIXTH UNITED STATES ARMY: REPORT OF THE LUZOIN CAMPAIGN: 9 JANUARY 194 – N30 JUNE 1945, VOLUME I*, 88.

cipitous ravines made all movements exceedingly difficult. The enemy had, moreover, utilized the terrain to best advantage by constructing numerous mutually supporting cave positions, which had to be reduced one by one, in order to permit the eastward advance of the Division to continue. The advance was, moreover, flanked 1,500-2,000 yards north of and parallel to the Villa Verde Trail by Mt. Imugan, on the forward slopes of which the enemy had established defensive positions and artillery observation posts. The Mt. Imugan positions dominated a stretch of over two miles of the Villa Verde Trail and his observation stations enabled the enabled the enemy to adjust his artillery fire on troops and vehicles moving along the trail, which ran along the crest of razor back ridges and formed the only route of advance. Besides, the Mt. Imugan positions enabled the enemy to repulse any direct attack through the valley north of the trail and constituted an ever-present threat to the line of communications of the 32nd Division. Under the circumstances, with the enemy holding Mt. Imugan, the 32nd Division had no choice but to crack the enemy defenses on the dominating hills directly in its front some four miles west of Imugan village, since bypassing them was impossible. The resulting struggle was slow and bloody and demanded the utmost of valor and fortitude on the part of our troops, especially since the division was unable to bring all its power into play, because it had to protect its rearward communications all the way from its front lines to Saint Nicholas.[176]

176 The 32nd "Red Arrow" Veteran Association, *The 32nd Infantry Division In World War II, The "Red Arrow" Luzon Campaign – The Villa Verde Trail*, (www.32nd-

General Gill echoed this description:

> ... the 32nd Division was headed for the Villa Verde Trail, which was nothing more or less than a foot trail over the mountains. As far as our 32nd Division was concerned, our first job was to build some kind of a road ... We had to supply ourselves with food and water, and while the road was being built we were fighting at the same time ... General Krueger came up to see me quite often on the Villa Verde Trail (once we got road enough for a jeep to come up) and he says in his book of having made several personal inspections, that his conclusion very definitely was that it would be a long, slow, and costly operation. And indeed that's what it turned out to be. Morale was poor because the men were tired. They had been in there in combat for months, they had had only a little rest after Leyte and during January, February, march, and April we had to fight the Japs on the trail and build the road at the same time.[177]

The Division AAR also noted:

> The VILLA VERDE TRAIL writhes over the harsh, rugged landscape, winding around and between high conical peaks which are separated by deep chasm-like ravines. Before the battle the slopes and peaks were covered with the secondary growth of the rain forest, which under artillery pounding became a matted, tangled mass. For the sake of clarity the peaks or hills in the SALACSAC PASS area were numbered. From west to

---

division.org/), Accessed 11/17/2024, 2.

177 Edward Jaquelin Smith, *Always a Commander: The Reminiscences of Major General William Gill*, 84.

> east from 502, with certain omissions, to HILL 533. SALACSAC PASS proper is the stretch of the VILLA VERDE TRAIL between Hills 504 and 508. Entrance to the PASS from the west is between Hills 504 and 505. This was designated SALACSAC PASS #2. The name SALACSAC PASS #1 was assigned to the gap through which the trail ram between Hill 508 to the south and the unnumbered hill to the north. Another mammoth ridge jutting generally west from MT IMUGAN and paralleling the VILLA VERDE TRAIL some 1,000 yards north of it, came to be known as YAMASHITA RIDGE. From the vastness of this ridge, the enemy harassed our troops along the trail with long-range machine gun and heavy mortar and artillery fire.[178]

Sixth Army knew what lay ahead and also knew that a minimum of 110,000 Japanese troops were in the north and only three divisions available to launch any type of offensive.[179] The Sixth Army report on the campaign further reinforces the land to be fought on and the enemy defending it:

> In an area pock marked with a multitude of cave positions so constructed as completely to dominate the trail and the valley to the north. The fact that the crest of a hull had been secured was no indication whatever that the hill itself was in our hands, for it was still necessary to undertake tedious mopping up operations to clean out or seal the numerous cave positions which covered the slopes o the hill. Moreover, outflanking an ene-

178 32d Infantry Division, *After Action Report: MIKE I Operation, Luzon, Jan. 27 – 30 June, 1945*, (Headquarters: 32d Infantry Division, 1945), 20.

179 Sixth Army, *SIXTH UNITED STATES ARMY: REPORT OF THE LUZOIN CAMPAIGN: 9 JANUARY 194 – N30 JUNE 1945, VOLUME I*, 57.

> my defensive position did not mean that it would collapse, for the enemy fought on in his caves and tunnels until they were blown up or sealed or their occupants killed. He fed reinforcements into each position as long as there was any way to reach it, no matter how untenable it might have become in the meantime. When the position became isolated, the Japanese defenders fought until overcome and killed.[180]

This was destined to be a hard fight and beyond the capabilities of the 32nd in its then current state. After the war General Gill lamented: "It's pretty hard to tell you now, after all these years, just how I felt during the battle on the Villa Verde Trail in Luzon, beginning in January and lasting until the end of the summer. But it was a fact that the 32nd Division was given a mission that was too big for its ability. No question in my mind."[181] They would fight on at great cost.

Once all division elements landed the infantry regiments were scattered about. The 126th initially remained around the landing beaches at Lingayen Gulf while the 127th and 128th IR's relieved elements of the 25th Infantry Division. The 127th occupied San Manuel, the 128th Assignan. The 2nd Bn of the 127th would be the unit leading the movement towards the Villa Verde Trail on 31 January. High ground dominated their advance and would do so throughout the campaign. On 5 February a preparatory artillery and mortar barrage preceded a 2nd Bn attack that gained 1,000 yards. That would seem extraordinary when compared to the coming months, but the battalion was "fighting up hill all the way, which was to be the case for many days to come..."[182] 6 February saw another gain

180 Ibid., 88.

181 Edward Jaquelin Smith, *Always a Commander: The Reminiscences of Major General William Gill*, 87.

182 32d Infantry Division, *After Action Report: MIKE I Operation, Luzon, Jan. 27 – 30 June, 1945*, 6.

of 1,500 yards; 8 February 2,500 yards.[183] While impressive there was a cost. The battalion was tiring, as noted by the Division AAR: "By now troops of this battalion were nearing exhaustion. The bare razorback ridge, around which the trail twisted and turned, left the troops exposed to the debilitating effects of the tropic sun. Water had to be carried from some distance back on the trail up the precipitous side of the ridge to the men on the crests."[184] The battalion fought off a Japanese night attack on 8/9 February and an attempt to resume the attack was stalled on the morning of the 9th by heavy small arms fire. On 10 February the 127th IR's 3rd Bn, thoroughly exhausted, relieved the 2nd Bn.

> **It was obvious by 10 February that the 2nd Bn, 127th Inf. Had been fighting through an outpost system organized in depth and that a major defensive position had now been encountered. This conclusion was borne out by the great volume of defensive fires coming from a series of ridges and knolls which commanded the VILLA VERDE TRAIL and which actually formed a series of mutually supporting "forts."**[185]
>
> —32nd Infantry Division Intelligence Assessment, 11 February 1945

The Division called the forthcoming fight the Battle of the Bowl. This "Bowl" was a series of knolls designated as the 3 series (3, 3A, 3B, 3C and 3D) and 5 series (5, 5A). This battle began with several days of vigorous patrolling to pinpoint the exact locations of enemy positions. Many units moved at night to thwart the Japanese's "unrestricted observation of all approaches to his positions."[186] The cliffs of these knolls made attacks difficult due to their near vertical

183 Ibid., 7.

184 Ibid.

185 Ibid., 9.

186 Ibid.

faces. Attacks were hard to coordinate because of the vegetation; one infantry company became isolated and ran into an enemy force double its size. They fought their way out but received no support because Japanese sniper fire kept reinforcements away.

This fight was followed by a week, 14-21 February, of American patrolling during the day and Japanese patrolling at night. On 22 February the 3rd Bn, 127th Inf. launched an attack. By 24 February the "Bowl" was captured. Tactical air power in the form of P-51 Mustangs through repeated bombing and strafing runs did not do any real damage to the enemy. However, it did keep them buried in their holes and caves and allowed U.S. troops to overrun their positions. Now came mopping up, the dreaded part of combat against the Japanese: "... mopping up was proceeding apace. This mopping up proved more costly to our troops than the main battle had. Enemy losses for the three-day period amounted to a counted 350 dead. The bitterness of the fighting is attested to by the fact that not one Jap prisoner was taken."[187] This is also where the physical and psychological deterioration of the Division began. They would consistently secure objectives and then fight just as hard once those objectives were "secure." Nothing ever seemed to end.

The 126th Inf. would run into heavy opposition after securing Sapinit, a small hamlet in the Ambayabang River Valley, on 28 February. As they continued moving forward their route became a gigantic S, the perfect site for multiple ambushes by a defensive force. The Division AAR noted:

> From the air the riverbed of AMBAYABANG appears as a gigantic S, deeply etched into the mountain range. Straight-line distance from serif to serif is no more than 2,000 yards. However, following the curves of the S through the gorge the distance is twice that. Here the Japs chose to make another delaying stand. That part of the riverbed connec-

187 Ibid., 11.

> ting the upper and lower hooks of the S offered an unobstructed field of fire for nearly a thousand yards. Along this stretch, the riverbed is 150 yards wide. Then too, the enemy dug his holes and caves in the steep walls of the gorge, taking advantage of the hull mass formed by a high ridge like nose, which protruded, into the lower hook of the S for protection from our artillery.[188]

The fight was unusual. 126th Inf. patrols made it to one of the serifs and noticed an enemy patrol beneath them. 12 Japanese soldiers, in a single file, were spotted walking along the riverbed. 11 American riflemen then in a textbook use of precision rifle fire eliminated all of them.[189] While that was a masterful, quick engagement, on the morning of 3 March it was back to the ruggedness typical of these battles. The Japanese were in their caves and it fell to the flamethrower and demolition teams to either burn or blow them out. It took time and slowed the advance. The terrain again gave the Japanese the advantage:

> ... The ruggedness of this terrain allowed a few defenders to slow to a snail pace the advance of our much stronger attacking force, while exacting a dear price for the ground yielded. To secure the valley for its supply route, the 3rd Bn, 126th Inf., was forced to clear the ridges on either side. While one company was eliminating the enemy on the ridge west of the valley, elements of the battalion had pushed ahead on the east ridge to contain the Jap force further north. This battalion was engaged in the slow process of sweeping the ridges until 10 March, when it was relieved by the 1st Bn. In this treeless terrain where tactical surprise was impos-

188 Ibid., 12-13.

189 Ibid., 13.

> sible and small unit maneuvering severely restricted by the precipitous slopes, our attacks followed a familiar pattern. Enemy positions were hammered unmercifully by the might of our artillery, after which the riflemen tried to close in on the Japs, still dazed and crouching at the bottom of their holes. **Failing as often as succeeding**, these small-scale attacks tried the courage and stamina of infantrymen to the straining point..."[190]

Earlier, I noted that the Japanese strength in northern Luzon far outnumbered that of the U.S. I Corps. It did but it was scattered throughout the northern part of the island. U.S. forces could and usually did have numerical superiority at the point of attack. However, given the terrain smaller enemy forces could frustrate attacks that should be quickly successful for days or even weeks. That's what happened when a single Japanese company, assisted by a reinforced platoon, as they tried to secure a ridge approximately two miles northeast of Ansagan, stalled 126th Infantry's 1st Battalion for an entire month.[191] The battalion would then need several weeks to rest and refit. This happened throughout this battle. All this was the preliminary. The main fight for the Villa Verde Trail began in earnest in March.

190 Ibid., 14.

191 Ibid., 18.

CHAPTER IX

THE VILLA VERDE TRAIL, PART II

> Forces on the VILLA VERDE TRAIL remained in contact with Japanese defenders, making slight gains. It was becoming apparent that a long siege would be required to dislodge the enemy from his successive commanding ridges along the trail. Maneuver was becoming increasingly difficult, the fighting ranging from valley floors up precipitous slopes to narrow mountain summits. At every turn of the winding trail forces found the enemy overlooking them from higher ground.
>
> —*I Corps: History Of Luzon Campaign, Philippine Islands 1945*

Robert Ross Smith wrote the Army's official history of the Luzon and Southern Philippines campaigns, *Triumph in the Philippines*. It is THE primary source. With the possible exception of the 32nd's official history of the campaign as far as tactical detail, there is nothing better. Smith interviewed everyone and had access to files and records that are not available today. He understood this campaign and in the following words describes what was now breaking a proud division:

> Whatever its shortcomings in other fields, the Japanese Army always had a feel for terrain, exploiting to the full every advantage the ground offered. Thus, as it moved up, the Japanese 2d Tank Division set to work to establish a system of mutually supporting defensive positions in order to control

every twist of the Villa Verde Trail and every fold in the ground throughout the pass area. Every knoll and hillock on or near the trail was the site of at least one machine gun emplacement; every wooded draw providing a route for outflanking a position was zeroed in for artillery or mortars. The cave, natural or man-made, came to characterize the defenses. Artillery was employed in quantity and quality not often encountered in engagements against the Japanese, who, as usual, made excellent use of their light and medium mortars. Finally, the 2d Tank Division was overstocked in automatic weapons, evidently having available many more than the 32d Division could bring to bear.

Against such defenses the 32d Division's difficult operations in the Salacsac Pass area could hardly avoid taking on a monotonous pattern. First, there would be unsuccessful frontal attacks against hillside strongholds. Failing, the troops would wait for air and artillery support to soften up the opposition and try again. Then there would be company and battalion outflanking maneuvers, some successful, some ending in near disaster, and all, as the result of Japanese defensive dispositions, inevitably winding up as frontal assaults. Every type of action would be repeated day after dreary day, either in heat enervating to the extreme on clear days, or in cloudbursts, fog, and mud. The nights were cold and, as the rainy season approached, increasingly damp and wet.

To reduce Japanese cave positions, the 32d Division would necessarily have to attack at least two mutually supporting caves simultaneously, at the same time endeavoring to keep flanking defensive installations neutralized by machine gun and mortar fire. Advances would depend upon a series of closely coordinated platoon actions, with platoons providing fire support for each other while each attacked its own objectives. Each cave, once neutralized, would have to be sealed; each position of other types would have to be occupied or the job would have to be done over and over again. Day after day units would have to patrol in order to locate routes to outflank known Japanese positions, ascertain Japanese flanks and pinpoint Japanese defenses for air and artillery bombardments.

In brief, the battle for the Villa Verde Trail became a knockdown, drag-out slugfest. The spectacular could hardly happen – there wasn't room enough. Troops would become tired and dispirited; nonbattle casualties would exceed those injured in combat. Supply would be very difficult, the evacuation of the sick and wounded an even greater problem. This was combined mountain and tropical warfare at its worst. The 32d Division had already had plenty of both, from the jungles of New Guinea to the mountains of western Leyte.[192]

192 Robert Ross Smith, *Triumph in the Philippines*, 497-8.

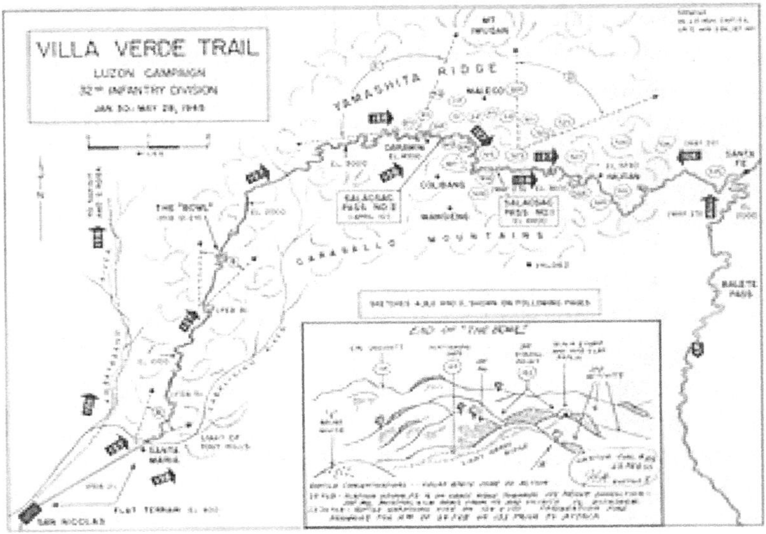

FIGURE XXVI. The "Hills" on the Villa Verde Trail

The above Hills dominated the next four months of the 32nd Infantry Division's life. Hill after Hill. The fighting was vicious, exhausting, and produced heavy casualties.

The above illustration dominated the operations of the 32nd until the end of May. Those numbered hills and Salacsac Pass #'s 1 and 2 were all protracted fights. From 8 – 14 March I Corps described the fights as follows:

> 32d Division forces continued to meet savage, well-organized resistance from Japanese dug in on the Villa Verde Trail. While the 1st Bn. Of the 127th IR continued attacking east from positions 4,200 yards west of IMUGAN making only slight gains, the 3rd Bn. Of the 127th on 8 March struck to the north from positions in VALDEZ and drove a wedge to the rear of the Japanese opposing the advance of the 1st Bn. From positions on high ground astride the Trail 3,000 yards west of IMUGAN, elements of the 3d Bn. Drove west in the hills sou-

> th of the Trail in an attempt to make contact with right flank elements of the 1st Bn. Remainder of the 3d Bn. held their ground tom protect the rear. The well-organized Japanese responded quickly to every thrust from the west as well as from the east with heavy artillery, machine gun, mortar and small arms fore, and in addition attempted numerous infiltration and banzai attacks each night. Fog and rain were another handicap to our forces. By the end of the period the 1st Bn. had advanced east to positions west of IMUGAN against continuous opposition while the 3d Bn. in the hills south of the Trail occupied commanding terrain 3,800 yards due west of IMUGAN but had failed to contact the 1st Bn. Japanese resistance remained at a high level.[193]

The 127th was the regiment that saw the most initial action. Even moving into position to attack was a laborious process as machine gun fire from the hills could hit anywhere on the Trail. Air strikes by two squadrons of P-51 Mustangs bombed the Hill, followed by artillery fire, and then the Mustangs returned for strafing runs. These preparatory fires allowed the infantry to get to the foot of the hill. But, what lay ahead did not garner confidence. The bombing and artillery fire had blasted jungle everywhere. Not only would it make an advance even harder it also made the Japanese and their cave positions more difficult to spot. The slope was also imposing, "... so steep was the climb that the troops of Company C had to scale the slope on all fours."[194] Once there one of the Company's platoons passed a series of Japanese cave positions but had no idea of it until they were fired on from the rear. Friendly mortar fire covered them as they fought back to the rest of the Company. The

193 I Corps, *History of Luzon Campaign, Philippine Islands, 1945*, 64.

194 32d Infantry Division, *After Action Report: MIKE I Operation, Luzon, Jan. 27 – 30 June, 1945*, 22.

FIGURE XXVII. Aerial photograph of the Villa Verde Trail

The above was the terrain of the Villa Verde Trail. Nowhere could the Japanese not observe the 32nd's advance. Its never ending winds and turns also were ripe for ambush.

Battalion Commander, who observed the attack saw that the air strikes, mortar and artillery fire, "... had little if any effect on the Japs inside the caves."[195]

After several days the 1st Bn. reached the top of Hill 502. Immediately, they were hit by machine gun fire from Hill 503. That was typical of this struggle. "Securing" one hill meant nothing. They were all within a few hundred yards of each other. The Japanese could bring machine gun and artillery fire on the 32nd at any time. Patrols also noticed Japanese reinforcements move to Hill 504. Another tough fight to think about. But first the dreaded mopping up of Hill 502 had to happen. The Division AAR does an outstanding job of describing this process:

195 Ibid.

> The laborious and costly process of reducing the cave positions is best understood by the movement of Company C for the next two days. Troops of this Company found, after several failures; that to reduce the cave defense system it was necessary to attack two or more positions simultaneously while keeping those on the flanks neutralized by machine gun and mortar fire. In this manner two platoons inched their way from the top of Hill 502 down two noses, which were separated by a ravine. These two platoons provided cross supporting fire for each other, while the heavy weapons company, from positions at the bottom of the hill, protected the flanks. Once on top of the caves, various methods were used to deal with the occupants. Smoke grenades, when thrown into an entrance revealed that the caves were connected by tunnels, since smoke would often come out of another hole some distance away. This led our troops to seal the entrances by caving in the ceiling with demolitions and by piling soil in front from above. Jap reactions to this tactic provided some uneasy moments to our troops. After sealing the caves during the day, our troops were kept awake during the night by the muffled sounds of the Japs trying to dig their way to the surface of the ground in the middle of a company perimeter. More often, though, Japs chose suicide when confronted with suffocation. The dull, faint thud of an exploding grenade coming from the inside of the sealed cave was usually heard by our troops after completing their dangerous and revolting task.[196]

This fighting saw more acts of bravery from the men of the 32nd.

---

196 Ibid., 29.

Two of the four Medals of Honor the Division received on Luzon were awarded for actions during this time period.

The first of these Medals of Honor went to Private First Class Thomas Eugene Atkins of the 127th Infantry Regiment:

> *... Private Atkins fought gallantly on the Villa Verde Trail ... With two companions he occupied a position on a ridge outside the perimeter defense established by the 1st Platoon on a high hill. At about 3 A.M. two companies of Japanese attacked with rifle and machine gun fire, grenades, TNT charges, and land mines, severely wounding PFC Atkins and killing his new companions. Despite the intense hostile fire and pain from his deep wound, he held his ground and returned heavy fire. After the attack was repulsed, he remained in his precarious position to repulse any subsequent assaults instead of returning to the American lines for medical treatment. The Japanese repeatedly made fierce attacks but for four hours, PFC Atkins determinedly remained in his foxhole, bearing the brunt of each assault and maintaining steady and accurate fire until each charge was repulse. At 7 A.M. 13 enemy dead lay in front of his position, he had fired 400 rounds, all he an his two companions possessed, an had used three rifles until each had jammed too badly for future operation. He withdrew during a lull to secure a rifle and more ammunition, and was persuade to remain for medical treatment. While waiting, he saw a Japanese within the perimeter and, seizing a nearby rifle killed him. A few minutes later, while lying on a litter, he discovered an enemy group, moving up behind the platoon's lines. Despite his severe wound, he sat up, delivered heavy rifle fire against the group and forced them to withdraw ...*[197]

197 https://www.32nd-division.org/old/history/moh/32moh.html, Accessed

The second Medal of Honor was awarded to Staff Sergeant Ysmael R. Villegas of the 127th Infantry Regiment for actions taken on 20 March 1945:

> ***... SSG Villegas was a squad leader when his unit, in a forward position, clashed with an enemy strongly entrenched in connected caves and foxholes on commanding ground. He moved boldly from man to man, in the face of bursting grenades and demolition charges, through heavy machine gun and rifle fire, to bolster the spirit of his comrades. Inspired by his gallantry, his men pressed forward to the crest of the hill. Numerous enemy riflemen, refusing to flee, continued firing from their foxholes. SSG Villegas, with complete disregard for his own safety and the bullets which kicked up the dirt at his feet, charged an enemy position, and firing at point-blank range killed the Japanese in a foxhole. He rushed a second foxhole while bullets missed him by inches, an killed one more of the enemy. In rapid succession he charged a third, a fourth, a fifth foxhole, each time destroying the enemy within. The fire against him increased in intensity, but he pressed onward to attack a sixth position. As he neared his goal. He was hit and killed by enemy fire.***[198]

The 127th Infantry Regiment would continue fighting in this constricted area. Between Hills 507 and 508 the Regiment tried to advance through a heavily wooded valley. Progress was slow as trails had to be hacked through the jungle. Heavy rains and low clouds also hampered movement. The weather also contributed to a serious problem that will be addressed a bit later, non-battle casualties. Sickness and exhaustion from the environment further reduced

10 March 2025.

198 https://www.32nd-division.org/old/history/moh/32moh.html, Accessed 10 March 2025.

the 127th's strength. While moving towards Salacsac Pass, around Hill 504, the 1st Bn. launched an attack that was stalled by heavy rain and fog, as well as Japanese mortar and machine gun fire. Not until 21 March was Hill 507 finally taken. The 127th was then driven off the hill that same day. While 507 was taken the hills around it were not. Machine gun and artillery fire pounded 507 and an enemy counterattack followed. Mercifully, the 127th IR was pulled off the line on 23 March, replaced by the 128th IR.[199] The situation which the 128th found itself in was: "Generally, the line which the 128th Inf. With its 1st Bn. took over fro the 127th Inf. Ran around the reverse slope of Hill 502, down into the saddle, the up to the crest of Hill 504 and bent back down the slope of the trail. We also held a foothold on Hill 505 . . . Enemy machine gun fire interdicted the trail between Hills 505 and 504, SALACSAC PASS #2, and also between Hills 504 and 502.[200] Rather than summarize and paraphrase I will quote the tactical situation as told by the Division AAR as well as the remedy applied by the 128th IR:

> *The reverse or northeastern slope of Hill 504 from the crest to the bottom was pock-marked with caves and tunnel openings. From positions on top and on the sides of Hill 503, Jap machine gunners protected the north flank of Hill 504. The south flank of Hill 504 was covered by automatic weapons fire from Hills 505, 506 and 507. Consequently, whenever our attacking troops attempted to go over the crest or around the sides of Hill 504 to get at the Jap positions on the reverse slope they ran into a veritable sheathe of bullets.*
>
> *Within each "hedgehog" position the principle of mutually supporting fires was adhered to consistently. Cave entrances and foxholes were so located, general-*

199 32d Infantry Division, *After Action Report: MIKE I Operation, Luzon, Jan. 27 – 30 June, 1945*, 24-26.

200 Ibid., 27.

ly in the walls of the vertical ravines which slashed the slopes, that each in turn protected and was protected by several others.

Hill 504, and the other hills as well, bore, a remarkable resemblance to a tremendous ant-hill, dotted by numerous entrances and honey-combed with a mass of tunnels. The enemy was not hesitant about directing machine gun fore or even artillery fire into his own positions when attackers penetrated into the system. Offset tunnels protected the occupants from friendly fire as well as from hostile bullets and demolitions.

Understandably, this ingeniously conceived interlocking defense system presented a dismaying problem to attackers, who were fired upon every time they tried to approach an emplacement. There seemed to be no starting point, no weak link in the chain. As it evolved, the solution to mutually supporting defenses was mutually supporting attacks. The experience of Company C, 127th Infantry, proved that caves which supported each other had to be attacked simultaneously by small units firing across the fronts of each other to keep the mutually supporting emplacements neutralized, that is, to keep all the Japs in that immediate area in their holes at the same time.

This principle was applied eventually by the 128th Infantry in reducing the strong points on Hills 504 and 505. As the 3d Battalion attacked over the crest and around the slopes of Hill 504, the 2d Battalion from its foothold on Hill 505 furnished neutralizing fires on the Jap positions on the lower slope of Hill 504 as well as Hill 506. Simultaneously the 3d Battalion poured its fire on the crest and sides of Hill 505 to protect the 2d Battalion as it pushed over the crest.

> *Direct wire communication between the two battalions facilitated the coordination of these supporting fires. Prodigious artillery and mortar concentrations were laid down on Hills 503, 506, 506A and 507 to keep the occupants of these positions concerned with their own safety. The enemy replied in kind, causing more casualties in our ranks with his artillery and mortar fire than with his small arms.*[201]

As described above, this process—while seemingly effective—required an immense amount of coordination and a prodigious use of artillery fires. U.S. Army artillery was a constant throughout the campaign. For the first time the 32nd encountered heavy doses of Japanese artillery fire. Japan occupied Luzon for three years. They had massed guns and ammunition during that time. Further, all the hills on the Villa Verde Trail were marked for their guns. Nowhere was American infantry safe. It would be the end of May before the Japanese guns were finally silenced.

27 March saw the 128th IR's 1st Bn. move east from Hill 504 towards Hill 505. The immediate objective was to "clean out" the area between the hills. A single Company had gained a foothold on Hill 505 and the rest of the battalion fought ferociously to join them. They succeeded. Yet even with success came problems. Not all the Japanese caves on Hill 504 had been destroyed or even identified. As the 1st Bn. moved onto Hill 505 they were subjected to enemy machine gun fire from their rear in addition to artillery, mortar, an machine gun fire emanating from 505. For five days, 28 March to 1 April, the 1st Bn., until relieved by the Regiment's 2nd Bn., worked to expand their foothold on Hill 505. Within close proximity to several caves the Battalion could do little as machine gun fire came from everywhere. On the night of 31 March Hill 504 was hit first by a heavy enemy artillery barrage and then a "mad attack" (as called by the Division) by several hundred infan-

201 Ibid., 28.

try. This attack drove the one company occupying 504 off. A rare night counterattack, primarily using grenades (an incredible 1,800 of them) due to fears of friendly casualties in the dark from small arms fire, drove the Japanese away. U.S. Losses were 84 killed and wounded. Enemy losses were 61 dead. The disparity in casualties is misleading. U.S. losses were almost always lower than that of the Japanese. Those 61 enemy dead were those found on the forward slope of Hill 504. The enemy dead on the hill's reverse slope could not be counted since it was covered by machine gun fire from all the other hills.[202]

During this time the Division's third Medal of Honor was awarded to Private First Class William Ralph Shockley of the 128th Infantry Regiment. PFC Shockley:

> ***... was in position with his unit on a hill when the enemy, after a concentration of artillery fire, launched a counterattack. He maintained his position under intense enemy fire an urged his comrades to withdraw, saying that he would "remain to the end" to provide cover. Although he had to clear two stoppages which impeded the reloading of his weapon, he halted one enemy charge. Hostile troops then began moving in on his left flank, and he quickly shifted his gun to fire on them. knowing that the only route of escape was being cut off by the enemy, he ordered the remainder of his squad to withdraw to safety and deliberately remained at his post. He continued to fire until he was killed during an ensuing enemy charge.***[203]

As with so many Medal of Honor recipients, this Medal was awarded posthumously. Acts of bravery such as the above are perilous.

202 Ibid., 30-31.

203 https://www.32nd-division.org/old/history/moh/32moh.html, Accessed 14 march 2025.

The effort required in these battles did not result in appreciable gains. A few hills were "secured" but how far had the Division advanced up the Villa Verde? A few hundred yards and they were nowhere near their ultimate objective of the Balete Pass. It took the 128th's 2nd Bn. ten days to secure one of those wretched hills, 2-11 April, and during that time the Bn.:

1. Destroyed two Japanese machine gun companies
2. Killed 223 enemy soldiers
3. Sealed 137 caves
4. Destroyed 12 light machine guns and 13 heavy machine guns
5. Destroyed U.S. equipment undoubtedly captured when the Philippines fell in 1942 including BAR's, Thompson Sub-Machine Guns, and flame throwers.[204]

Again the number of Japanese dead was misleading. Most of those 137 caves were sealed. Few were explored. The Japanese dead or those left to die was never known.

The 126th Infantry Regiment was also having its problems. By 7 April, they were at the base of Hill 511 but could go no further. Artillery fire poured on them and for the first time the Division acknowledged a growing number of shock cases. Frustrated in its inability to advance the Regiment turned to sapper trenches. Two cave entrances were especially troublesome. Expertly located they were invulnerable to mortar and artillery fire. For eighteen hours everyone took a turn digging until the trenches were within only a few feet of the cave entrances. Blocks of TNT were then tossed into the caves. The enemy machine guns were captured and the caves sealed with sandbags in addition to the demolished cave entrance. Weeks later when all of Hill 505 was totally secure it was discovered:

204 Lt. Colonel Robert B. Vance. *Action On Hill 505*, (Headquarters: 1st Bn., 128th Infantry Regiment, 22 June 1945), 1-3.

> ... that Hill 505 had been so thoroughly excavated by the Japs that it was actually a shell. From a huge central cavern about 2o feet by 20 feet, tunnels radiated out in all directions and opened at various levels on slopes of the hill. Machine gun emplacements had been ingeniously contrived. Generally, emplacements were dug to conform to the outline of the weapon. Offset escape hatches led from the emplacements down into a chamber. During a barrage, the operator removed the gun and lowered it down the escape hatch into the chamber. When emplaced the gun fired along a final protective line. This allowed the operator to fire without exposing himself. Standing on a ladder in the escape hatch, the operator merely turned the elevating wheel until the muzzle of the gun cleared the top of the emplacement. In some cases operators were equipped with periscopes; in other cases an observer at a different position signaled to the operator to open fire.[205]

There were some two dozen numbered hills on the Trail in addition to many other small ones. Some were as well designed as Hill 505, some even more, some undoubtedly less. Each was a fight and there were not enough troops to handle them all. This leads to the question: why were there no more troops?

In mid-April, the 32nd noted that: "Despite the deep penetration our troops had made into his SALACSAC PASS redoubt, the enemy still showed no slackening in his determination to stand to the end. Indeed, his counterattacks became more vicious, though no more effective in regaining lost ground than previously."[206] However, it was effective in reducing American combat abilities and

205 32d Infantry Division, *After Action Report: MIKE I Operation, Luzon, Jan. 27 – 30 June 1945*, 36.

206 Ibid., 37.

increasing physical and psychological exhaustion. Reinforcements and replacements should not be confused as one in the same. Reinforcements bolster an organization's numbers. Replacements step in for those lost. For the 32nd there would be no reinforcements. Replacements seldom measured up to those they stood in for.

The 127th IR was shattered when the 128th IR moved in to relieve it on 23 March. After a few weeks it would return to action with only 2,600 men well below their authorized strength.[207] A month later the 128th was, as noted by campaign historian Robert Ross Smith, "finished as a fighting unit."[208] The Regiment had 710 battle casualties during that month and was also down to 1,500 men. Losses in the 126th IR, while not as heavy, 70 dead and 145 wounded, still reduced their total number to 2,100.[209]

Those above the Division noted the lack of progress of the 32nd ID. General Krueger sent observers to the Trail and they reported back to the Sixth Army commander:

1. The 126th and 128th Infantry Regiments were approaching complete mental and physical exhaustion.
2. Front-line men with considerable time overseas were becoming super cautious.
3. Rotation back to the United States had become the principal topic of conversation at all echelons of the division.
4. The combat troops' aggressive spirit was diminishing rapidly and markedly.[210]

None of the above should be surprising. One-third of the Division had been there since Buna. They had then fought along the Driniumor River, on Leyte, and in the hopeless jungles and hills of the Villa Verde Trail. They had been in combat more than any other

207 Robert Ross Smith, *Triumph in the Philippines*, 502.

208 Ibid., 503.

209 Ibid.

210 Ibid.

American infantry division of the Second World War. They were of course tired. Even if the average infantryman of the 32nd did not see the uselessness of the campaign itself (as I pointed out in a previous chapter), they could see two things: the possible end of the war or the invasion of Japan. That could be both inspiring and depressing. Leaders were also harder to find, especially from the Company level down to even the standard rifle squad. Morale began to suffer, it had to. Replacements were needed to restore ranks but reinforcements were needed far more.

General Gill saw all this happening. He had to be frustrated and after a conversation with General Innis Swift, his Corps Commander, helpless:

> Along about the first of April I began to realize just how bad off we were, so I asked Major Gen. Swift, commander of the 1st Corps, what we could do about getting any reinforcements. His reply was also very definite: there were no reinforcements available either in the corps or in General Krueger's army. There was very small chance of getting anything and I would have to use what I had to accomplish my mission. It's all well and good to sit back and say that ... But there are casualties other than battle casualties, particularly after a long period of hard fighting and exposure to the elements, to the diseases and to all the things that go to reduce the capabilities of fighting men. For example, we had all sorts of dysentery and malaria cases, the weather was bad ... Those things contributed a great deal to the lessening of our fighting abilities.[211]

---

211 Edward Jaquelin Smith, *Always a Commander: The Reminiscences of Major General William Gill*, 87.

How did this happen? How could no forces be available? Again, we return to MacArthur. To him, it was as if the fighting on Luzon ended with the capture of Manila. A Captain from the 128th IR, Robert Maynard, stated that: "MacArthur took Manila ... then he didn't care about the dying in the mountains."[212] In addition to wanting all of Luzon taken, condemning the 32nd as well as the other divisions of the I Corps to six months of useless misery, he wanted all the other Japanese occupied islands in the vast Filipino island chain recaptured as well. Cebu, Panay, Negros, and Mindanao were isolated after the Leyte landings. Five divisions: the 24th, 31st, 40th, and 41st Infantry Divisions, and the Americal (Infantry) Division, were taken from Sixth Army to create the Eighth Army under General Eichelberger. Approximately 100,000 men that could have bolstered not only the 32nd but also the other I Corps units. Other divisions were out there but MacArthur had them off the line, preparing for an invasion of Japan. Of course pulling the I Corps back or allowing them to form a defensive line never entered MacArthur's mind. The 32nd had to continue to fight with what they had.

With no reinforcements available and depleted ranks, to even function the Division needed replacements. Unfortunately, replacements when they arrived were not on par with those they stood in for. Those coming from Replacement Depots were sent right to units already in combat. No time to acclimate, no time for jungle training, no time to assimilate into units and learn from the veterans. The other replacements were members of the Division returning from stints in the hospital for minor wounds or physical and mental exhaustion. They never returned better than they left.

The issue of non-battle casualties was both new and a thing of the past. During World War I it was called shell shock. During World War II it was referred to as battle fatigue. Today we call it Post Traumatic Stress Disorder (PTSD). I have thoroughly examined

212 World War II (Unknown Author), "The U.S. 32nd Infantry Division Battle to Control the Villa Verde Trail," *World War II Magazine*, June 12, 2006, 1.

the records of the 32nd ID as well as those of dozens of other divisions and their subordinate units. The issue of non-battle casualties is rarely even mentioned. While doing this research I found the Division's 127th IR and Robert Ross Smith's *Triumph in the Philippines* as the only thorough explanations of this problem. I am most impressed by the 127th's Surgeon, Major Louis A. Sass. His report is insightful and I therefore include several pages of it below:

> *The psychiatric problem within the 1st Bn. during the period 17 April to 27 April requires some explanation. During this period the 1st Bn. took the brunt of active combat and one would expect that the men would be under more nervous strain than either of the other two battalions, particularly when they were recipients of concentrated enemy artillery and mortar fire, counter-attacks an bombing by our own planes. It is noted that of the 268 hospitalized, 34 cases or 12.6% were from nervous disorders. Some of these men had shown symptoms of anxiety state, neuro-circulatory asthenia and emotional instability previous to combat. The majority of these cases however, were the direct result of the stresses of frontline combat in infantry companies.*
>
> *In examining the possible causes for these psychiatric sick, it is important to point out the circumstances attending the situation. The regiment had been in combat for almost two months and had been returned to a rear area for a three-week rest period before returning to the Trail for the second time. It was during this rest period that replacements who had just arrived overseas, and had never been subject to combat, joined the companies. The majority of men were 18-20 years of age, an age where even in civilian life governments do not consider them sufficiently stable emotionally to entrust them with adult decisions and situations.*

Shortly after, the 1st Bn. was committed to fight in exceedingly difficult terrain against an enemy who were using more artillery and mortar than we had ever previously encountered. Under such conditions the recent replacements were subjected to a heavy punishment for which they had not been psychologically prepared. The brief stay they had had with the company before entering combat was insufficient for them to become psychologically attached and develop an attitude of loyalty and comradeship ... The willingness to share the dangers with their fellow soldiers and the unwillingness to leave these comrades had not been developed. Not all of these psychiatric patients were recent replacements, a substantial number of men had seen previous combat. Here the additional mental strain of numerous campaigns, previously being wounded, long years overseas, lack of adequate rest periods between campaigns, etc., played a part. In addition to heavy artillery and mortar fire, there occurred the unfortunate experience of having our air force drop several bombs on our own front line, burying a number of men and wounding others. This unfortunate incident contributed to the demoralization of some of the troops.

The symptoms exhibited by these casualties were varied but certain signs and symptoms were common to all. They exhibited extreme nervousness, anxiety, generalized weakness, some weeping, depression and in a few cases, hysteric conversion symptoms. About 50% were placed on phenobarbital and returned to duty. A few adjusted and were able to perform efficiently. The remainder cowered in their foxholes, trembling and in fear that they were unable to control until they were returned to the aid station by their officers as of no use to the company and a demoralizing

> *influence on the other men in the line. It was to such cases that the diagnosis of combat fatigue or exhaustion, instead of psychoneurosis, was attached in accordance with directives from higher echelons cautioning against the use of the diagnosis of psychoneurosis in combat and thus fix in the patient's mind that was "psycho" or mentally ill. The connation of "fatigue" and "exhaustion" denotes in the mind of the patient that quiet and rest lead to complete recovery.*
>
> *It is felt that a good many of these patients evacuated, especially among the new replacements, could have been returned to duty in a week or less by rear echelons, particularly the clearing station. It is hear that men should be given rest, well-prepared hot meals, sedation and psychotherapy. Instead, patients were immediately evacuated from the clearing station to the rear, with the result that it was weeks later before they rejoined their unit through replacement channels, giving them additional time to meditate their worries . . .*[213]

Dr. Sass was a remarkable person. Unfortunately, there were few like him. The doctors in combat units were brave and did remarkable work. There were just not enough of them. Further, those that were there rarely had any psychological training. The Army did not seem to want to recognize the psychological effects of sustained combat without significant periods of rest. Men within these units also did not willingly acknowledge either to themselves, their superiors, or medical personnel that something was wrong. Perhaps veterans would feel like they would be abandoning those they had served with for years. New replacement troops perhaps did not want to look weak or scared before veterans. Yet no one can blame a teenager of 18 or 19 years old from the Midwest if after a few

213 Major Louis A. Sass, *Surgeon's Report, Luzon Campaign,* (Headquarters: 127th Infantry Regiment, 1945), 7-10.

days in a jungle with the overbearing heat, incessant rains, constant artillery fire, the noises of the jungle, and little sleep they became a psychological casualty. Those who did report their problem would be sent to a hospital, in most cases, for only a few days. The doctors at the hospitals or aid stations had commanders screaming when will they get their men back. As time wore on, the Division's problem was not just a lack of numbers. The men, through no fault of their own, were not what they once were.

The 128th IR was at the same point the 127th IR had been when they replaced them on 23 March. The 127th, after several weeks of rest, returned to relieve the 128th; they began arriving on 18 April. They noted that "[a]s the latter moved out, they showed considerable evidence of battle fatigue demonstrated by short temper and general harassed feeling throughout."[214] The 127th seemed reinvigorated after their rest. On 19 April the Regiment notes: "To date the morale of the troops is good to excellent. The status of all our troops is amazing, they hold up and fight like fury in the face of tremendous odds as to tactical position."[215] After only a few days the Regiment then noted that: "Morale on 21, 22, and 23 April is becoming questionable in a few instances. There are numerous cases of psychoneurosis, manifested by extreme nervousness and acute fear. These cases are at present occurring among the new replacements. The older men, particularly, those who have been with the regiment since 1942, are extremely tired. It will not seem strange if some, at the least, of the latter group show signs of psychoneurosis in the future."[216] Sixteen days after their relief the 128th returned to action, the 127th pilled back once again.[217] The cycle would continue until the end of May. The 126th IR was not replaced when it

214 127th Infantry Regiment, History of the 127th Infantry Regiment: Luzon Campaign: 27 January 1945 – 30 June 1945, (Headquarters: 127th Infantry Regiment, 1945), 3.

215 Ibid., 5.

216 Ibid., 9.

217 128th Infantry Regiment, *Historical Report: Period 27 January 1945 to 30 June 1945*, (Headquarters: 128th Infantry, 32nd Infantry Division, 1945), 16.

pulled back. A few weeks of rest and right back into the fight.

Despite this debilitating situation there were still remarkable acts of bravery. The Division's fourth Medal of Honor was awarded to Private First Class David Maldanado Gonzales of the 127th IR for his actions on 25 April 1945:

> ***... PFC Gonzales was pinned down with his company. As enemy fire swept the area, making any movement extremely hazardous, a 500-pound bomb smashed into the company's perimeter, burying five men with its explosion. PFC Gonzales, without hesitation, seized an entrenching tool and under a hail of fire crawled 15 yards to his entombed comrades, where his commanding officer, who had also rushed forward, was beginning to dig the men out. Nearing his goal, he saw the officer struck and instantly killed by machine gun fire. Undismayed, he set to work swiftly and surely with his hands and the entrenching tool while enemy sniper and machine gun bullets struck all about him. He succeeded in digging one of the men out of the pile of rock and sand. To dig faster he stood up regardless of the greater danger from so exposing himself. He extricated a second man, and then another. As he completed the liberation of the third, he was hit and mortally wounded, but the comrades for whom he so gallantly gave his life were safely evacuated.***[218]

There were others that deserved multiple awards. These men just happened to be the ones recognized as individuals.

John Carlisle was a reporter who accompanied the 32nd in what he termed, its "119 Grueling Days" on the Villa Verde Trail. One of

218 https://www.32nd-division.org/old/history/moh/32moh.html, Accessed 15 April 2025.

the many fights he described occurred in May but was typical of the combat that continued until the end of that month:

> The troops have spotted a Jap cave with perhaps 25 Japs in it. The caves support each other with machine gun fire. We of the assault platoon want to get within 15 feet of that cave to let it have a charge. So we draw on friendly troops for support.
>
> The riflemen get as close as they can, up to 25 or 35 yards from the cave. We move in, crawling, clawing our way up ahead of them. Automatic riflemen are a few feet behind us.
>
> Now an assault team is made up of two men, one carrying a charge of 30 pounds of TNT, the other a 15-foot bamboo pole. The charge is put on a C ration board rigged to the pole, with a primacord to set off the charge, and dynamic caps to detonate it.
>
> While we are crawling up to the cave, machine guns and riflemen fire at the cave. Others pick off nearby snipers.
>
> We get up into position, a few feet from the cave. We lay down and start pushing the charge in. When I start the fuse, we have 15 seconds to get out of there, or be blown up. We have to pull out fast—and we do.
>
> And what happens after the charge goes off?
>
> It seals the cave—we hope. Usually it does. The tunnel's cave in, and the entrance is closed forever. Sometimes we see smoke and dirt coming out above the cave. Then we know there is another entrance to seal. We go and get it.[219]

---

219 John M. Carlisle, *Red Arrow Men: Stories About the 32nd Division and the Villa*

So the fighting continued.

Between 17 April and 4 May, with the 128th off the line, getting another quick rest, the 126th and 127th IR's continued the frustrating campaign. Losses continued to mount. For the 126th there were 56 killed and 173 wounded, the 127th had 128 killed and 353 wounded. In addition, 503 more non-battle casualties added to the growing total.[220] But, finally, the Japanese resistance was crumbling. By the end of May the fighting along the Trail was essentially over. The Division never did reach Santa Fe. In four months they had advanced approximately 25 miles. In those 25 miles they had officially accounted for 5,750 of the 8,750 men of the Japanese 2d Tank Division.[221] Most of the rest were buried in the hundreds of sealed caves in the never-ending hills of the Trail or had withdrawn further north into the mountains. Final Division casualty numbers can fluctuate by a small number between various accounts, but Robert Ross Smith's appears to this author the most accurate as he acknowledges the non-battle casualties:

	Killed	**Wounded**	**Total**
126th Infantry Regiment	195	460	655
127th Infantry Regiment	350	750	1,100
128th Infantry Regiment	280	950	1,230
	825	2,160	2,985[222]
Total Division Non-battle Casualties: 6,000			

The saddest part of the above numbers is strategically, they were killed, wounded, or evacuated for no reason. Luzon was militarily secure. The mountainous north, of which the Villa Verde Trail was a part, should have been left isolated. MacArthur had done that before in New Guinea, why not here? However, the bravery and

Verde, (Nashville: The Battery Press (Reprint), 1990) 64.

220 Robert Ross Smith, *Triumph In The Philippines*, 507.

221 Ibid., 510.

222 Ibid., 509.

devotion to duty of the men, of this Division, can NEVER be forgotten. They fought on and on. General Gill at the end of the campaign acknowledged their valor:

> *The 32d Division has accomplished its mission. The enemy had been destroyed and the Villa Verde Trail secured. A passage has been forced through the Caraballo Mountains from the Central Plain to the entrance of the Cagayan Valley, thus hastening the completion of the Luzon Campaign.*
>
> *After one hundred and twenty days of fierce hand to hand combat over terrain more difficult than any yet encountered in this war, the "Red Arrow" again pierced the enemy's line. You have crushed completely another of the enemy's so-called impregnable defenses, brilliantly concluding the Division's 5th Campaign in the Pacific Theater.*
>
> *I desire to express to every officer and enlisted man in the Division, as well as those attached, my heartfelt appreciation of the courage and determination each has shown while playing his vitally important role in this long and arduous campaign. You have outfought and destroyed a cunning and determined enemy, an enemy occupying elaborately prepared defenses on ground of his own choosing. Your victory was impressive and decisive and one of which you may well be proud. It is with justifiable pride and complete confidence that I look forward to your continued success into the heart of Tokyo.*[223]

They would not go to Tokyo. Fortunately for the 32nd Infantry Division the war was now over.

223 https://www.32nd-division.org/?s=general+gill+message, Accessed 25 April 1945.

FIGURE XXVIII. 32nd Infantry Division Memorial on the Villa Verde Trail

CHAPTER X

THE FACILITATORS

The great majority of this work focuses on the infantry. It was the infantry, the ordinary foot soldiers, that fought the Japanese face to face throughout the war. They suffered most of the casualties. Yet for them to be able to fight they needed the help of the Division's other organic parts as well as various attachments and outside support. While I do not dedicate enough to their own sacrifices and accomplishments, I will attempt to acknowledge their impact.

Administrative military services are often deemphasized or even completely neglected. As the war progressed from Buna to the end of operations on Luzon the Division's administrative section took on greater responsibilities. Personnel issues are the main task of the unit administrators. At Buna and in the months before those issues were a tall task as the 32nd took on thousands of new soldiers while at the same time phasing many out. Once the battle began the Division at no time received reinforcements but men were pulled off the line as both battle and non-battle casualties and many returned. After Buna, administration facilitated between February 1943 and February 1944 nearly 7,000 new men for not only the infantry but all other needed specialties. On Luzon, in a difficult environment that saw no break in combat operations, 238 officers and 6,661 enlisted men entered the Division as replacements while 214 officers and 5,747 enlisted men returned after hospitalization.[224] As stated by the Division's Veterans Association: "... the overall figures indicate that a tremendous personnel administrative and training burden was carried by the Division during an exceptionally difficult period of combat."

224 https://www.32nd-division.org/, Accessed 28 April 2025.

An army not properly supplied cannot win in combat. At Buna supply was an enormous problem. The United States was not yet truly mobilized and as described in the Buna chapter, there were shortages of everything from heavy weapons to food. Transport, by land, sea, and air was sparse. The victory at Buna given the dire supply situation is a testament to the bravery and determination of the men who fought there despite these shortages. Inadequate supply did not affect the Division for the rest of the war. There was enough food, water, ammunition, artillery and every other necessity of war. The problem became getting it to the troops. The PTO was not the ETO. The Division fought on New Guinea, at the time still one of the most primitive places on Earth. There were no roads. Fortunately, after Buna the fighting along the Driniumor was near the coast. But even getting the supplies forward usually meant carrying them on the backs of soldiers or native laborers. That meant small arms ammunition, mortar ammunition, food, and water (the most vital item in the jungle climate), all heavy items, could run short. When the units of the Driniumor Covering Force initially broke it was not because they ran out of ammunition. There were just too few men covering too large an area. From New Guinea the 32nd moved to Leyte. While not a modern road network, even in 1944, it was at least useable. Artillery ammunition became vitally important and the Division supply sections kept the guns firing, tens of thousands of rounds. On Luzon, the Villa Verde Trail must have made those with the 32nd since Buna think they returned to that misery. I have repeatedly described that the so-called Trail was barely that. Getting supplies forward, under fire from the Trail's maze of hills, was dangerous but yet again these men did their part. The Division was fighting the Japanese on the Villa Verde Trail while also trying to build a supply road. Here the Division engineers stood up.

Engineers, who refer to themselves as Combat Engineers, are a vital component of modern war. The Division's engineers did not deploy to Buna with the infantry regiments. Most of the Division support elements were left behind because the transport to move them forward was non-existent. On the Driniumor they made the positions occupied far stouter and helped the infantry fight against an attack where at some points they were outnumbered 10 to 1. On Leyte they built roadblocks, destroyed other ones, and improved roads. On Luzon, they built a road in the mud and jungle while receiving sniper, machine gun, and artillery fire. That road kept the Division moving and fed. Without the Combat Engineers the ordeal of the Villa Verde Trail would have been far worse. It was not without cost as many an engineer vehicle slid off embankments or was ambushed in the jungle.

Artillery is often referred to as the "Queen of Battle." Due to the unavailability of tanks because of terrain limitations and tactical air support because of its ineffectiveness it certainly was the Queen to the infantry's King. At Buna there was one American gun, a single 105-mm. howitzer. It performed valuably once it received the ammunition it needed. There was a shortage of infantry along the Driniumor but not artillery. When the Japanese came screaming out of the jungle towards the American line preregistered targets were blanketed by artillery fire. Thousands of enemy troops fell. On Leyte they again performed excellent service but it was on Luzon where they played a truly decisive role. Because of the formidable Japanese positions and their resistance to standard infantry weapons, the Division artillery fired 6,000 missions and expended 200,000 rounds of 105-mm. and 155-mm. ammunition.[225] As the Artillery AAR notes:

> It is believed that no artillery unit has yet been called upon, in this war, to deliver accurate, continuous close support under more unfavorable and difficult conditions ... 200,000 rounds of accurate, timely, and effective fire upon the enemy and his formidable defenses, thus giving that close continuous support tom our infantry that enabled the

225 32nd Division Artillery, *Report After Action: M-1 Operation, 27 January 45 to 30 June 45,* (Headquarters: 32nd Division Artillery, 32nd Infantry Division, 1945), 1.

> RED ARROW DIVISION, in the longest and hardest fight of its history, to crus completely another of the enemies so called impregnable defenses.[226]

Another weapon could have been invaluable to the infantry: tanks. But, terrain restrictions limited their value.

At Buna a small number of light tanks turned the tide. On the Driniumor the forward elements had a detachment of tank destroyers in support but the terrain took away their mobility. They became pillboxes. On Leyte, in small numbers, they supported advances and eliminated Japanese positions. They could have made a difference on the Villa Verde but the Trail was impossible for a 30-ton tank to traverse. Their value was underrated. Air power, on the other hand, was vastly overrated.

Buna demonstrated how ineffective tactical air power was in a jungle. Targets were impossible to see and the close proximity of American and Japanese troops made any air attack as dangerous to U.S. troops as it was to the enemy. Nowhere were the differences between fighting in Europe and the Pacific starker. The ETO did see tragic mistakes. July 1944's Operation *COBRA* saw two examples of inaccurate bombing causing heavy American casualties. Those mistakes were made in saturation bombing runs by strategic bombers in a tactical support role. Heavy bombers never conducted such a mission in the 32nd's four major campaigns. Tactical air support was decisive in North Africa and Western Europe. On New Guinea and the Philippines it was nonexistent. Air power also became necessary at times to drop supplies because of the lack of roads. Yet that was risky since "... low flying clouds will make air-dropping impossible at times and inaccurate on the best of days."[227] Just another weapon rendered ineffective by the environment of the PTO.

---

226 Ibid., 2.

227 27th Infantry Regiment, *History of the 127th Infantry Regiment: Luzon Campaign: 27 January 1945 – 30 June 1945*, 4.

Finally, where would the Division have been without its medical personnel? From the medics to the doctors, they played a valuable part in all the 32nd campaigns. Medics, right at the front of the battle, took care of the wounded and the doctors like Major Cass tried to get them ready to return to duty or for evacuation. At Buna, so early in the war where medical supplies were as limited as everything else, their work was truly remarkable. Hundreds of killed and wounded and thousands of non-battle casualties when even less were aware of what combat related psychological problems actually entailed. Till the end of the Luzon campaign, they did what they considered their most important task, taking care of the welfare of the men.

THE WAR IS OVER ...

When thinking of the famous divisions of World War II, does anyone ever mention the 32nd Infantry Division? In the Pacific the famous divisions are all in the Marine Corps. The Army divisions fought in Europe. The Army is ignored for its role in the victory over Japan. NO American division spent as many days in combat during World War II as did the 32nd Infantry. The bravery and perseverance of this division is remarkable. It is also sad that they started in the misery of Buna and ended with the misery of the Villa Verde Trail. They deserved better than that. Fortunately the war ended for them and for so many other infantry divisions with the dropping of the atomic bombs. What possibly lay ahead, had the bombs not ended the war, was an invasion of Japan. The 32nd Infantry Division was preparing to be one of the units MacArthur would lead, from the safety of a naval vessel far offshore, onto Japan itself. Forecasts for potential American casualties were staggering. Instead, the Red Arrow men finally went home.

APPENDIX I

DISTINGUISHED SERVICE CROSSES AT BUNA

In addition to 2 Medals of Honor awarded in this battle, the 32nd Infantry Division was awarded 100 Distinguished Service Crosses (99 American, 1 Australian), the nation's second highest military award. It received 157 for the entire war, two-thirds given in this, their first battle. It represents the extraordinary bravery exhibited by the men who fought here. Below are the names of those conferred with the DSC:

Major Bernard G. Baetcke, 30 November 1942

1st Lieutenant Cladie Alford Bailey, 2 December 1942

Private First Class Walter A, Bajdek, 10 December 1942

Private First Class William J. Balza, 25 December 1942

Major Chester Martin Beaver, 5 December 1942

Private First Class Herman Bender, 31 December 1942

Private First Class Jack M. Binns, 11 December 1942

Captain William Francis Boice, 19 December 1942

Captain Herman Johann Bottcher, 5-11 December 1942 and 20 December 1942

Captain Joseph Sladen Bradley, 28 December 1942

Private First Class Leonard Taylor Bradley, 5 December 1942

Brigadier General Clovis Ethelbert Byers, 15 December 1942

Private Robert H. Campbell, 17 December 1942 – 1 January 1943

Colonel John Joseph Carew, 16 November 1942

Staff Sergeant Carl J. Cherney, 20-21 November 1942

First Lieutenant Claude Ellis Clark Jr., 19-21 November 1942

First Lieutenant James T. Coker, 8 January 1943
Private John E. Combs, 1 December 1943
First Lieutenant John William Crow, 19-20 November 1942
Private First Class Jack K. Cunningham, 11 January 1943
First Lieutenant Peter Louis Dal Ponte, 9 December 1942
Staff Sergeant Delmar H. Daniels, 1 December 1942
Technician Fifth Grade Edwin C. De Rosier, 19 November 1942
Second Lieutenant James D. Doughtie, 23 December 1942
First Lieutenant James Garnett Downer, 9 December 1942
Sergeant Frank David Duket Jr., 24 December 1942
Private Howard M. Eastwood, 26 November 1942
Private Gordon Wayne Eoff, 25 December 1942
Private Bernardino Y. Estrada, 16 December 1942
Sergeant William F. Fale, 25 December 1942
Private First Class Albert L. Fisher, 24 December 1942
First Lieutenant Powell Alexander Fraser, 11 January 1942
Second Lieutenant Talley Doyle Fullmer, 31 December 1942 – 11 January 1943
Sergeant Chester C. Punk, 23-24 December 1942
Private Harold E. Graber, 5 December 1942
Technician Fifth Grade Charles H. Gray, 28 December 1942
Captain Millard G. Gray, 25 December 1942 – 1 January 1943
Colonel John E. Grose, 2 January 1943
Private Elmer R. Hangartner, 28 December 1942
Captain Harold Elwood Hantelmann, 1-3 December 1942
First Lieutenant John E. Harbert, 16 November 1942
Lieutenant Colonel Merle Henry Howe, 5 December 1942
Captain Meredith Miller Huggins, 1 December 1942
First Lieutenant James I. Hunt, 2-5 December 1942

Private Earl W. Johnson, 23-31 December 1942

Private Maro P. Johnson, 6 November 1942

Private Raymond R. Judd, 12 January 1943

First Lieutenant Thomas E. Knode

Corporal Delos A. Leland, 19 November 1942

Sergeant Boyd L. Lincoln, 30 November 1942

Second Lieutenant Paul R. Lutjens, 5 December 1942

Staff Sergeant John R. MacGowan, 16 November 1942

Lieutenant Colonel Alexander J. MacNab, 10 December 1942 – 3 January 1943

Brigadier General Hanford MacNider, 23 November 1942

Private Lawrence B. Marion, 24 December 1942

Colonel Clarence A. Martin, 3 December 1942 – 5 January 1943

Second Lieutenant Frederick W. Matz, 24 December 1942

Private Homer W. McAllister, 16 November 1942

First Lieutenant Robert P. McCampbell, 27 December 1942

Lieutenant Colonel Melvin Leslie McCreary, 24 December 1942

Technician Fifth Grade Bart T. McDonough, 28 December 1942

Sergeant Robert R. McGee, 23 – 28 November 1942

Private Arthur J. Melanson Jr., 28 December 1942

Private first Class Raymond Milby, 12 January 1943

Staff Sergeant Milan John Miljatovich, 28 December 1942

Corporal Harold L. Mitchell, 7 – 9 December 1942

Private First Class Earl Mittleberger, 28 December 1942

Private Cloyd G. Myers, 16 November 1942

First Lieutenant Erwin Joseph Nummer, 30 November 1942

Private Steve W. Parks, 20 December 1942

First Lieutenant Herbert Gade Peabody, 16 November 1942 and 5 December 1942

Private Harold O. Pederson, 24 December 1942

Private Marvin M. Peterson, 12 January 1943

Private First Class Donald M. Price, 16 November 1942

Staff Sergeant John F. Rehak Jr., 20 December 1942

Corporal Daniel F. Rini, 16 December 1942

Private First Class Edward R. Rossman, 24 December 1942

Captain Donald Frank Runnoe, 2 January 1942

Major Edmund R. Schroeder, 27 December 1942

Second Lieutenant Paul L. Schwartz, 5 December 1942

Staff Sergeant Herman Tate Shaw, 25 December 1942 – 8 January 1943

Major Herbert M. Smith, 7 December 1942

Corporal Gordon C. Snyder, 16 November 1942

Private Lawrence F. Sprague, 12 January 1943

Private Edward G. Squires, 28 December 1942

First Sergeant Reuben J. Steger, 21 November 1942

Corporal Orrin C. Sutton, 24 December 1942

Second Lieutenant John Emmett Sweet, Jr., 26 December 1942

Staff Sergeant Robert George Thompson, 11 January 1943

Sergeant Francis J. Vondracek, 24 December 1942

Brigadier General Albert Whitney Waldron, 5 December 1942

Major Simon Warmenhoven, 26 November 1942

Sergeant Howard J. Weiss, 16 November 1942

Captain Russell J. Widley, 26 December 1942

Sergeant Samuel G. Winzenried, 13 December 1942

Captain James W. Workman, 24 December 1942

Major Roy E. Zinser, 16 December 1942

Below are several write-ups of these individual acts of heroism. Most of these are chosen because of the articles printed in Wisconsin local newspapers, which were more detailed than the citations themselves. There are several which provide the exact citation.

Private First Class William J. Balza was a member of Headquarters Company, 2nd Battalion, 127th Infantry Regiment. On 26 January 1943, Robert J. Doyle of *The Milwaukee Journal* wrote:

> ***PVT William (Balza), Route 4 Sawyer, dared death three times near Buna mission Christmas day. The third time he succeeded in his mission and has received a Distinguished Cross. Japanese soldiers allowed a company to cross an open field, then cut the company off from its unit by sweeping the field with machine gun fire every time anyone attempted to reach the isolated company. (Balza) volunteered to brave this fire to string a telephone line across this field. Twice he and SGT Elmer Coon, Black Creek, WI, wormed into knee high grass but were forced back. The second time a soldier trying to protect them with rifle fire was killed. (Balza) volunteered to make a third attempt and finally succeeded as machine gun bullets cut the grass all around him. The wire which he dragged across the field maintained communication between the isolated company and other units until help was brought.***

Private First Class Herman Bender was a member of Company M, 127th Infantry Regiment. Robert J. Doyle of *The Milwaukee Journal* once again highlighted an American hero on 26 June 1943:

> ***A Distinguished Service Cross was awarded posthumously to Pvt. Herman Bender, Fond du lac, who gave his life getting important information for his commanding officer. Bender volunteered to cross nearly 200 yards of unpatrolled ground to attempt***

to make contact between two American forces closing on the enemy beachhead. He reached the other force, obtained necessary information and had started back to report when he was severely wounded. Bender dragged himself to his platoon and made his report shortly before he died. The information was of great and enabled the two American forces to close the gap for the final attack against Japanese positions at Buna Mission.

Colonel Joseph Sladen Bradley was a member of the 32nd Infantry Division's Headquarters Section. His citation read:

... for extraordinary heroism in connection with military operations against an armed enemy while serving as Chief of Staff, Headquarters, 32d Infantry Division, in action against enemy forces on 28 December 1942m during the Papuan Campaign at Buna, New Guinea. To further an attack by an infantry company against a strong fortified enemy area, Colonel Bradley left his command post, went to the most forward positions, personally aided the progress of the attack, and materially to its success. While under heavy enemy fire he moved about among the men continually risking his life to encourage and assist the advance in every way possible. Colonel Bradley's inspiring leadership, personal bravery and zealous devotion to duty exemplify the highest traditions of the military forces of the United States and reflect great credit upon himself, the 32d Infantry Division, and the United States Army.

First Lieutenant John E. Herbert was a member of Headquarters Company, 32nd Infantry Division. His citation read:

... for extraordinary heroism in connection with military operations against an armed enemy while

serving with Headquarters, 32d Infantry Division, in action against enemy forces on 16 November 1942, near Cape Sudeat, New Guinea. When enemy airplanes attacked and set fire to a small ship, First Lieutenant Herbert organized a party of men to pull to shore a barge loaded with men and ammunition. Throughout repeated strafing by hostile planes, he refused to take cover and continued his efforts. When flaming debris from explosions of munitions aboard the ship fell on the barge, he remained aboard and personally assisted in throwing the fragments overboard and in extinguishing the flames. His courage and calmness set a heroic example to men who had sought cover and so steadied and inspired them that work was resumed and the unloading of the badly needed ammunition on the barge was successfully accomplished ...

Captain Meredith Huggins was a member of Headquarters, 3rd Battalion, 126th Infantry Regiment. This account comes from the December 1943 issue of *Phi Gamma Delta*:

For extraordinary heroism in action near Sananan-da, New Guinea, on 1 December 1942. Captain Huggins successfully led a party through enemy-infested terrain to the assistance of an isolated force which was holding an important position behind the enemy lines. Shortly after the arrival of his party, a strong attack by the enemy was repulsed. During this action the commanding officer was killed. Captain Huggins assumed command and expertly deployed his men to the best advantage. In so doing he exposed himself to heavy enemy fire and sustained a gunshot wound. Although his force was completely surrounded by the enemy. Captain Huggins inspired his men by the example of his courage and by skillful tactics succee-

> ***ded in holding the position until relief was effected. His courageous leadership saved a most precarious situation.***

First Lieutenant James I. Hunt was a member of Headquarters Company, 126th Infantry Regiment. His citation read:

> ***... for extraordinary heroism in connection with military operations against an armed enemy while serving with an Anti-Tank Platoon of Headquarters Company, 126th Infantry Regiment, 32d Infantry Division, in action against enemy forces from 2 to 5 December 1942, during the Papuan Campaign at Buna Village, First Lieutenant Hunt, when the attack was checked by hostile fire, crawled forward without regard to his personal safety and threw a hand grenade into an enemy machine gun emplacement. Although wounded above the knee during this action, he continued to lead his platoon in the attack. Evacuated two days later because of his wound, he voluntarily returned to duty with his unit the following day, although the bullet had not been removed and his hospitalization was recommended.***

Second Lieutenant was a member of Company L, 127th Infantry Regiment. On 26 June 1943 Robert Doyle of *The Milwaukee Journal* wrote:

> ***When his mortar platoon assisting riflemen in an attack was halted by enemy fire, Matz advanced alone and knocked out the enemy pillboxes. Though wounded he continued to press forward. Wounded again when the enemy counterattacked, Matz sent back his men and remained with the soldier unable to walk. Matz and the other soldier were given up for dead but reached their own lines Jan. 2 when Buna Mission fell. Matz has recovered from his wounds.***

Private Marvin Peterson was a member of Company C, 127th Infantry Regiment. On 3 July 1943 Robert Doyle again wrote in *The Milwaukee Journal*:

> ***Peterson was part of a force advancing from the Buna area along the coast against the last strong Japanese positions in Papua. The force had to cross the swift Konombi river Jan. 21 to dislodge enemy positions on the opposite bank. Peterson and three others volunteered to swim the river in broad daylight with enemy guns trained on them. Wearing only shorts and carrying pistols and hand grenades, the four swam the rain swollen river and fastened a rope which enabled a platoon to cross and establish a bridgehead.***

Sergeant Francis J. Vondracek was a member of Company I, 127th Infantry Regiment. On 26 January 1943 Robert Doyle wrote in *The Milwaukee Journal*:

> ***SGT Francis Vondracek, Route 2, Kaukauna, led a platoon in an attack near Buna Mission Dec. 24 and has received a DSC for heroism. Heavy fire stopped the platoon and Vondracek reported back to his commanding officer, but when his men were withdrawn Vondracek volunteered to try to knock out three pillboxes which had pinned his unit down all day. Armed with four grenades and with three riflemen covering his advance, Vondracek crawled within six feet of the first pillbox. He hurled four grenades and knocked out the pillbox, killing five Japs inside. Vondracek then moved back, filled his helmet with grenades and attacked two more pillboxes. He reached a shell hole near the pillboxes and despite heavy enemy fire hurled his grenades so accurately into the entrances that he knocked out both pillboxes.***

These are but a few of the heroic actions taken by the soldiers that fought at Buna. There were, as there always are, many who received neither recognition nor award for those actions.

The source for all the above citations is the 32nd Infantry Division's Veterans Organization: http://www.32nd-division.org/association/join.html. The citations are boldened and italicized because that is how they appear on the Division website.

APPENDIX II

DISTINGUISHED SERVICE CROSSES IN NEW GUINEA, 1944

While the 32nd Infantry Division received two Medals of Honor for actions at the Driniumor River it received only one Distinguished Service Cross. One! It seems the heroism and bravery of this campaign was neglected as the men of the Covering Force were by their higher commands. For actions at Saidor, a very small operation that preceded the Hollandia and Aitape landings by other units of 6th Army, Division soldiers received five DSCs. It is hard to understand. The men that fought at Saidor, as brief a battle as it was, undoubtedly deserved their awards. So did more of the men along the Driniumor River. As with Appendix I, I will detail some of the citations in their entirety.

Staff Sergeant Joseph P. Bayeri, 7 July 1944 (Driniumor River)

Lieutenant Colonel Gordon Madison Clarkson, 28 January 1944 (Saidor)

First Lieutenant George Hess, 28 January 1944 (Saidor)

Private First Class Van William Hill, 28 January 1944 (Saidor)

Private First-Class John F. Wagner, 4 January 1944 (Saidor)

Corporal Bernard P. Woods, 28 January 1944 (Saidor)

Lt. Col. Gordon Madison Clarkson, Commanding Officer of the 1st Battalion, 128th Infantry Regiment, received his DSC for the following actions on 28 January 1944:

> ***Lt. Col. Clarkson personally led group of volunteers that rushed to aid 1st Lt. George J. Hess' group after it became isolated on north bank of Mot River near Sai-***

> *dor on 28 Jan. '44. The group was part of the 50-may patrol from 1ST Bn., 128TH Inf. that was attacked by a 100-man Japanese force west of the Mot River near Teterei, along the western edge of the Saidor beachhead, earlier that day. The rescuers formed a human chain reaching from the bank into the river to catch 1st Lt. Hess as he was trying to carry a rope across the raging river, so his men could attempt to get back to the friendly side. Clarkson was the anchor of the chain on the river bank and was WIA by machinegun fire while trying to pull Sgt. Thomas R. Reed up the bank to find cover from the enemy fire. Sgt. Reed then pulled Lt. Col. Clarkson into the grass for concealment but he DOW soon after.*

1LT George I. Hess received his DSC during the same action as Lt. Col. Clarkson, mentioned above. His citation reads:

> *First Lt. Hess earned the DSC for his actions as co-leader of 50-may patrol from 1ST Bn., 128TH Inf. that was attacked by a 100-man Japanese force west of the Mot River near Teterei, along the western edge of the Saidor beachhead, on 28 Jan. '44. The patrol split into three groups while it worked its way back to friendly lines on the east side of the river. His group was evacuating 1 WIA and 2 KIA and became isolated on north bank of Mot River. At least twice he was nearly swept away while trying to carry a rope across the raging river, so his men could attempt to get back to the friendly side. The patrol killed 43 Japanese soldiers, but suffered 8 killed, 6 MIA, and 9 WIA.*

APPENDIX III

LEYTE AWARDS, 127TH INFANTRY REGIMENT

WAR DEPARTMENT CITATION:

The 3^{D} BATTALION, 127TH INFANTRY REGIMENT, is cited for outstanding performance of duty in action against the enemy near Limon, Leyte, Philippine Islands, from 30 November to 7 December 1944. The 3^{D} BATTALION, 127TH INFANTRY REGIMENT, was ordered to attack Hill 400, near Limon, Leyte, Philippine Islands. This hill was the key defensive position of the crack Japanese Imperial 1st Division. The assigned mission was to knock out all installations, annihilate the foe, capture and hold commanding ground which he occupied. The commanding ground of Hill 400 enabled the Japanese not only to defend the hill proper but extensive areas on both the flanks as well. Automatic weapons, mortars, artillery, and small arms which the enemy possessed in great numbers could be utilized from these positions with maximum effectiveness. The irregularity of the slopes and dense undergrowth in many places was a definite handicap to any attacking force. In addition, for several hundred yards from the crest of the objective, the ground was completely barren, affording neither cover nor concealment from enemy observation and fire. The signal to attack was given. Company L on the right made a frontal attack; Company I, supported by Company K, advanced to make a left enveloping movement; Company M, stationed on Bridge Ridge, gave supporting 81mm, 60 mm mortar and overhead machine gun fire. For 8 days a vicious battle raged. Foot by foot, yard by yard, the men of the 3^{D} BATTALION pressed forward against fanatical enemy resistance. Numerous "banzai" counter attacks at times halted our men, and on occasion even compelled them to yield some ground, but their courage and determination never wavered. Temporary reverse only spurred them on to greater efforts. Individual acts of outstanding heroism were numerous.

Time and again officers and enlisted men, severely wounded, continued to fight on, rejecting medical aid and refusing to leave the field of battle. Headquarters company personnel kept a constant flow of ammunition and supplies to our embattled forces, despite the terrific enemy fire constantly directed at them. Medical officers and enlisted men attached to the battalion treated our wounded under fire and evacuated them promptly to the rear. These operations were carried out heedless of their personal safety and numerous enemy efforts to prevent their missions of rescue. On the eighth day our men were ready for the final assault on Hill 400. Two hundred yards of barren, open, and exposed ground lay between them and the crest of the hill. Across this ground, devoid of cover and concealment, gallantly our men braved the enemy's point-blank fire. Doggedly they advanced until the crest was reached. Hand to hand combat ensued when they attained the hill, but our men were not to be denied; they destroyed many defenders and forced the others to flee in panic. In the 8 day battle many enemy were killed and much valuable equipment captured or destroyed. A large number of maps and regimental personnel records of great intelligence value were also captured. The signal victory broke the enemy's claimed impregnable Yamashita Line, and contributed greatly to the complete and utter defeat of the Japanese troops resisting in the upper Ormoc Valley, and the final collapse of all enemy resistance on Leyte. In outmaneuvering, out fighting, and out lasting a numerically superior foe who had an overwhelming advantage in position and firepower, the officers and men of the 3D BATTALION, 127TH INFANTRY REGIMENT, exemplified the finest traditions of American arms."

DISTINGUISHED UNIT CITATION, COMPANY C, 127TH INFANTRY REGIMENT:

COMPANY C, 127TH INFANTRY REGIMENT, is cited for outstanding performance of duty in action against the enemy near the Ormoc Highway, Leyte, Philippine Islands, on 15 December 1944. As a result of a wide flanking movement by our forces a large Japanese pocket was trapped. The only supply or evacuation route open to this enemy

force was the Ormoc Highway. On 15 December 1944, COMPANY C ... was ordered to establish a roadblock on the Ormoc Highway to prevent these troops from being supplied or reinforced. Many enemy tanks were known to be in the area and were expected to resist any attempt by our forces to establish or maintain the block. The men of COMPANY C, armed with grenades, rifles, automatic weapons, mortars and bazookas, had just moved into position when the enemy struck. Ten light tanks supported by well-armed ground troops comprised the enemy force. In addition, the enemy directed intense 75-mm and 150-mm artillery fire against COMPANY C's positions. Although the men were subjected to point-blank fire from the enemy tanks, a hail of bullets from the numerically superior enemy ground troop, and the point-blank fire from the enemy artillery, they never ceased to face the challenge and return the fire to the foe. The men of COMPANY C, by accuracy of their fire and excellent coordination of small arms and bazookas, completely routed the enemy. The mortars and a special patrol succeeded in putting out of action two 75-mm and one 150-mm artillery piece which the enemy had been employing. When the smoke of the battle cleared there remained nine enemy tanks completely destroyed and the other severely damaged. The crews of these tanks as well as many of the supporting ground troops were killed. An aggressive assault against the disorganized remnants of the routed foe resulted in his complete annihilation. The courage and disregard for personal safety shown by the officers and enlisted men of COMPANY C, are a credit to the armed forces. This operation was a determining factor in the final successful break-through by our troops during the following days and played a vital role in the successful completion of the task of breaking all enemy resistance on Leyte.

The above citations are bolded and italicized because that is how they appeared in the history of this campaign on the 32nd Infantry Division Veteran's Organization website.

APPENDIX IV

DISTINGUISHED SERVICE CROSSES: LUZON

Staff Sergeant Henry Ellwood Brooks, 127th Infantry Regiment, 24 April 1945

Technical Sergeant George C. Budal, 128th Infantry Regiment, 8 May 1945

Lt. Colonel James P. Burns, 128th Infantry Regiment, 19 and 24 May

Corporal Arthur G. Clough, 128th Infantry Regiment, 31 March 1945

Captain Sheldon M. Dannelly, 127th Infantry Regiment, 4-5 March 1945

Major Timothy C. Doherty, 127th Infantry Regiment, 5-22 March 1945

Private First Class Henry Fernandez, 126th Infantry Regiment, 21 March 1945

1st Lieutenant Robert L. Fierman, 126th Infantry Regiment, 21 March 1945

Sergeant William C. Furr, 128th Infantry Regiment, 29 March 1945

Major General William Gill, 16 April – 15 May 1945

Colonel John A. Hettinger, 128th Infantry Regiment, 27 March 1945

Colonel Merle Henry Howe, 128th Infantry Regiment, 11 May 1945

Private First Class George E. Lamb, 128th Infantry Regiment, 28 March – 9 April 1945

Lt. Colonel Charles R. Meyer, 127th Infantry Regiment, 6 May 1945

Private First Class Donald W. Mullens, 126th Infantry Regiment, 21 March 1945

Private Calvin E. Ross, 127th Infantry Regiment, 15 & 20 March 1945

Technical Sergeant Learville N. Schlessman, 128th Infantry Regiment, 25 May 1945

Technical Sergeant Richard F. Serrahn, 127th Infantry Regiment, 10 March 1945

1st Lieutenant Robert J. Stallman, 127th Infantry Regiment, 18 March 1945

Captain Henry G. Taber, 127th Infantry Regiment, 9 February 1945

Staff Sergeant Robert Van Bogart, 127th Infantry Regiment, 5 March 1945

APPENDIX V

IMAGES FROM THE VILLA VERDE TRAIL

UNPUBLISHED SOURCES

Army Ground Forces. *Observers Reports w/ Appendixes: New Guinea*. Headquarters: Army Ground Forces, 1943.

Carroll, Lt. Colonel Leonard S. *Destruction of Japanese 1st Division in the Leyte Campaign*. Headquarters: United States Army Forces in the Far East, 22 February 1945.

Committee 16. *Armor On Leyte*. Fort Knox, Kentucky: The Advanced School, Officers Advanced Course, 1949.

Eichelberger, Robert. *Report of the Commanding General Buna Forces on the Buna Campaign, Dec. 1, 1942 – Jan. 25, 1943*. Headquarters: SWPA, 1943.

Eichelberger Diary. 1942.

Eichelberger to Dr. Samuel Milner, Personal Correspondence Concerning his book, Victory In Papua. 8 March 1954.

Handy, H.F., Colonel. *Observer Report: Southwest Pacific Theater of Operations, 23 December 1942*. Headquarters: Army Ground Forces, 1942.

I Corps. *History of the Buna Campaign: December 1, 1942 – January 25, 1942*. Headquarters: I Corps, 1943.

History of Luzon Campaign, Philippine Islands. Headquarters: I Corps, 1945.

Jones, Laurence M., Lt. Col. *Defense of the Philippines to the Battle of Buna: Critical Analysis of General Douglas MacArthur*. Maxwell Air Force Base, Alabama: Air War College, 1989.

Knight, Harry, Colonel. *Observer Report: Southwest Pacific Theater of Operations, 4 February 1943*. Headquarters: Army Ground Forces, 1943.

Marshall, General George C. *Comments on Combat Readiness*. Washington D.C.: Office of the Chief of Staff, 1944.

Military Intelligence Division. *The Buna-Sananda Operation, 16*

November 1942 – 23 January 1943. Washington D.C.: War Department, Military Intelligence Division, 1944.

Milner, Samuel. *Milner Interview with General Eichelberger.* 21 Nov 48.

Milner Interview with General Harding. 9 Dec 1947.

Milner interview/ General Waldron (Buna Artillery Commander). Undated.

Mueller, Charles J., Captain. *History Of The One Hundred Twenty Seventh Infantry Regiment: "Buna Campaign," 16 November 1942 to 1 February 1943*. Headquarters: 127th Infantry Regiment, 1943.

Samuel Milner Interview with General Harding, 9 Dec 1947.

Sass, Major Louis A. *Surgeon's Report: Luzon Campaign.* Headquarters: 127th Infantry Regiment, 1945.

Sixth Army. *SIXTH UNITED STATES ARMY: REPORT OF THE LUZON CAMPAIGN: 9 JANUARY 1945 – 30 JUNE 1945, VOLUME I*. Headquarters: Sixth Army, 1945.

Unknown Author, *Eclectic Memories and Opinions, 32nd Infantry Division, WWII, Southwest Pacific* Wisconsin: 32nd Veterans Association, 1946.

Vance, Lt. Colonel Robert B. *Action On Hill 505*. Headquarters: 1st Battalion, 128th Infantry Regiment, 1945.

32nd Division Artillery. *Report After Action: M-1 Operation, 27 January 45 to 30*

June 45. Headquarters: 32nd Division Artillery, 32nd Infantry Division, 1945.

32d Infantry Division. *After Action Report: Buna*. Headquarters: 32d Infantry Division, 1943.

After Action Report: MIKE I Operation, Luzon, Jan. 27 – 30 June, 1945. Headquarters: 32nd Infantry Division, 1945.

REPORT AFTER ACTION OF 32ND Infantry Division

in Leyte Engagement of K-2 Operation for the period 16 November – 25 December 1944. Headquarters: 32nd Infantry Division, 1945.

126th Infantry Regiment. *REPORT AFTER ACTION: LEYTE OPERATION*. Headquarters: 126th Infantry Regiment, 1945.

SUPPLEMENTARY REPORT AFTER ACTION, 26 Dec. 44 – 24 Jan. 45. Headquarters: 126th Infantry Regiment, 1945.

127th Infantry Regiment. *History – Leyte – 127th Infantry Regiment – 32nd Infantry Division, 6 November 1944 – 5 January 1945*. Headquarters: 127th Infantry Regiment, 1945.

History of the 127th Infantry Regiment: Luzon Campaign: 27 January 1945 – 30 June 1945. Headquarters: 127th Infantry Regiment, 1945.

128th Infantry Regiment. *Historical Report: Period 27 January 1945 to 30 June 1945*. Headquarters: 128th Infantry, 32nd Infantry Division, 1945.

HISTORY: LEYTE. Headquarters: 128th Infantry Regiment, 1945.

Report of Action Against Enemy: Buna. Headquarters: 128th Infantry Regiment, 1943.

PUBLISHED SOURCES

Cannon, M. Hamlin. Leyte: *The Return To The Philippines*. Washington D.C.: Center Of Military History, 1954.

Carlisle, John. *Red Arrow Men: Stories About the 32n Division and the Villa Verde*. Nashville: The Battery Press (Reprint), 1990.

Drea, Edward J. *Defending the Driniumor: Covering Force Operations in New Guinea 1944*. Fort Leavenworth, Kansas:

Combat Studies Institute, 1984.

MacArthur's ULTRA: Codebreaking and the War Against Japan, 1942-1945. Kansas: University of Kansas Press, 1992.

Eichelberger, Robert. *Our Jungle Road To Tokyo. New York*: The Viking Press, 1950.

Ellman, James. *MacArthur Reconsidered: General Douglas MacArthur as a Wartime Commander.* New York: Stackpole Books, 2023.

FM 17-32. *The Tank Company Light And Medium*, Headquarters: Department of the Army, 1942.

FM 17-36, *Employment of Tanks with Infantry*, Headquarters: Department of the Army, March 1944.

FM 72-20, *Jungle Warfare*, Headquarters: Department of the Army, October 1944.

FM 100-5, *Field Service Regulations (Operations)*, Department of the Army, 1942.

Kenney, George. *George Kenney Reports*. New York: Duell, Sloan, and Pearce, 1949.

Leary, William M. *We Shall Return: MacArthur's Commanders And The Defeat of Japan*. Kentucky: The University of Kentucky Press, 1988.

MacArthur, Douglas. *Reminiscences*. New York: Time, Inc., 1964.

Manchester, William. *American Caesar: Douglas MacArthur 1880–1964*. New York: Little, Brown, and Company, 1978.

Milner, Samuel. *Victory in Papua*. Washington D.C.: Office of the Chief of Military History, 1957.

Pogue, Forrest C. *Organizer of Victory, 1943–1945*. New York: Penguin Books, 1973.

Schaller, Michael. *Douglas MacArthur: The Far Eastern General.* London: Oxford University Press, 1989.

Smith, Edward Jaquelin. *Always A Commander – The Reminiscences of General William H. Gill*. Colorado: Colorado College, 1974.

Smith, Robert Ross. *The Approach To The Philippines*. Washington D.C.: Center of Military History, 1953.

Triumph in the Philippines. Washington D.C.: Center of Military History, 1963.

Van der Vat, Dan. *The Pacific Campaign: The U.S. – Japanese Naval War 1941–1945*. New York: Simon and Schuster, 1991.

Willoughby, Charles and Chamberlain, John. *MacArthur 1941–1951*. New York: Company, Inc. 1954.

Young Robert. *Pacific Hurtgen*. Washington D.C.: Westphalia Press, 2017.

ARTICLES

Lt. Colonel Joseph I Greene, Editor. *Infantry Journal*. 1942.

John Shortal. "Hollandia: A Training Victory." *Military Review*. May 1986.

Unknown Author. "The U.S. 32nd Infantry Division Battle to Control the Villa Verde Trail." *World War II Magazine*. June 12, 2006.

Related Titles from Westphalia Press

The Limits of Moderation: Jimmy Carter and the Ironies of American Liberalism

The Limits of Moderation: Jimmy Carter and the Ironies of American Liberalism is not a finished product. And yet, even in this unfinished stage, this book is a close and careful history of a short yet transformative period in American political history, when big changes were afoot.

The Zelensky Method
by Grant Farred

Locating Russian's war within a global context, The Zelensky Method is unsparing in its critique of those nations, who have refused to condemn Russia's invasion and are doing everything they can to prevent economic sanctions from being imposed on the Kremlin.

Sinking into the Honey Trap: The Case of the Israeli-Palestinian Conflict
by Daniel Bar-Tal, Barbara Doron, Translator

Sinking into the Honey Trap by Daniel Bar-Tal discusses how politics led Israel to advancing the occupation, and of the deterioration of democracy and morality that accelerates the growth of an authoritarian regime with nationalism and religiosity.

Essay on The Mysteries and the True Object of The Brotherhood of Freemasons
by Jason Williams

The third edition of Essai sur les mystères discusses Freemasonry's role as a society of symbolic philosophers who cultivate their minds, practice virtues, and engage in charity, and underscores the importance of brotherhood, morality, and goodwill.

Bunker Diplomacy: An Arab-American in the U.S. Foreign Service by Nabeel Khoury

After twenty-five years in the Foreign Service, Dr. Nabeel A. Khoury retired from the U.S. Department of State in 2013 with the rank of Minister Counselor. In his last overseas posting, Khoury served as deputy chief of mission at the U.S. embassy in Yemen (2004-2007).

Managing Challenges for the Flint Water Crisis Edited by Toyna E. Thornton, Andrew D. Williams, Katherine M. Simon, Jennifer F. Sklarew

This edited volume examines several public management and intergovernmental failures, with particular attention on social, political, and financial impacts. Understanding disaster meaning, even causality, is essential to the problem-solving process.

User-Centric Design by Dr. Diane Stottlemyer

User-centric strategy can improve by using tools to manage performance using specific techniques. User-centric design is based on and centered around the users. They are an essential part of the design process and should have a say in what they want and need from the application based on behavior and performance.

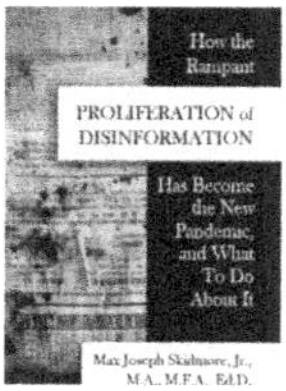

How the Rampant Proliferation of Disinformation Has Become the New Pandemic, and What To Do About It by Max Joseph Skidmore Jr.

This work examines the causes of the overwhelming tidal wave of fake news, misinformation, disinformation, and propaganda, and the increase in information illiteracy and mistrust in higher education and traditional, vetted news outlets that make fact-checking a priority.

Abortion and Informed Common Sense by Max J. Skidmore

The controversy over a woman's "right to choose," as opposed to the numerous "rights" that abortion opponents decide should be assumed to exist for "unborn children," has always struck me as incomplete. Two missing elements of the argument seems obvious, yet they remain almost completely overlooked

The Athenian Year Primer: Attic Time-Reckoning and the Julian Calendar
by Christopher Planeaux

The ability to translate ancient Athenian calendar references into precise Julian-Gregorian dates will not only assist Ancient Historians and Classicists to date numerous historical events with much greater accuracy but also aid epigraphists in the restorations of numerous Attic inscriptions.

Siddhartha: Life of the Buddha
by David L. Phillips,
contributions by Venerable Sitagu Sayadaw

Siddhartha: Life of the Buddha is an illustrated story for adults and children about the Buddha's birth, enlightenment and work for social justice. It includes illustrations from Pagan, Burma which are provided by Rev. Sitagu Sayadaw.

Growing Inequality: Bridging Complex Systems, Population Health, and Health Disparities
Editors: George A. Kaplan, Ana V. Diez Roux, Carl P. Simon, and Sandro Galea

Why is America's health is poorer than the health of other wealthy countries and why health inequities persist despite our efforts? In this book, researchers report on groundbreaking insights to simulate how these determinants come together to produce levels of population health and disparities and test new solutions.

Issues in Maritime Cyber Security
Edited by Dr. Joe DiRenzo III, Dr. Nicole K. Drumhiller, and Dr. Fred S. Roberts

The complexity of making MTS safe from cyber attack is daunting and the need for all stakeholders in both government (at all levels) and private industry to be involved in cyber security is more significant than ever as the use of the MTS continues to grow.

Female Emancipation and Masonic Membership: An Essential Collection
By Guillermo De Los Reyes Heredia

Female Emancipation and Masonic Membership: An Essential Combination is a collection of essays on Freemasonry and gender that promotes a transatlantic discussion of the study of the history of women and Freemasonry and their contribution in different countries.

Made in the USA
Monee, IL
04 November 2026

67383937R00128